CURRENT PERSPECTIVE ON IRRIGATION AND DRAINAGE

Edited by **Suren Kulshreshtha**
and **Amin Elshorbagy**

Current Perspective on Irrigation and Drainage

http://dx.doi.org/10.5772/63177
Edited by Suren Kulshreshtha and Amin Elshorbagy

Contributors

Joost Wellens, Bernard Tychon, Dirk Raes, Ernest Nnadi, Andy Shuttleworth, Fredrick Mbanaso, Stephen Coupe, Joris Voeten, Alan Newman, Toru Watanabe, Dong Duy Pham, Rousseau Tawegoum, Florian Leroy, Gérard Sintes, Gérard Chassériaux, Daniel G Bucur, Sjoerd E.A.T.M. Van Der Zee, Sija F. Stofberg

CBS, Edition **2018**

Published by InTech

Janeza Trdine 9, 51000 Rijeka, Croatia

Publishing Process Manager Romina Rovan

Technical Editor SPi Global

Cover Designer InTech Design Team

Additional hard copies can be obtained from orders@intechopen.com

Current Perspective on Irrigation and Drainage, Edited by Suren Kulshreshtha and Amin Elshorbagy
p. cm.
Print ISBN 978-953-51-2951-6
Online ISBN 978-953-51-2952-3

CURRENT PERSPECTIVE ON IRRIGATION AND DRAINAGE

Edited by **Suren Kulshreshtha** and **Amin Elshorbagy**

Current Perspective on Irrigation and Drainage

http://dx.doi.org/10.5772/63177
Edited by Suren Kulshreshtha and Amin Elshorbagy

Contributors

Joost Wellens, Bernard Tychon, Dirk Raes, Ernest Nnadi, Andy Shuttleworth, Fredrick Mbanaso, Stephen Coupe, Joris Voeten, Alan Newman, Toru Watanabe, Dong Duy Pham, Rousseau Tawegoum, Florian Leroy, Gérard Sintes, Gérard Chassériaux, Daniel G Bucur, Sjoerd E.A.T.M. Van Der Zee, Sija F. Stofberg

CBS, Edition **2018**

Published by InTech

Janeza Trdine 9, 51000 Rijeka, Croatia

Publishing Process Manager Romina Rovan

Technical Editor SPi Global

Cover Designer InTech Design Team

Additional hard copies can be obtained from orders@intechopen.com

Current Perspective on Irrigation and Drainage, Edited by Suren Kulshreshtha and Amin Elshorbagy
p. cm.
Print ISBN 978-953-51-2951-6
Online ISBN 978-953-51-2952-3

Contents

Preface

This book is a collection of chapters on various topics related to irrigation and drainage. The chapters contained in the book are the scholarly contributions of various authors pertinent to the science of irrigation and drainage. Each contribution is presented as a separate chapter complete in itself but directly related to the subject of this book.

This book contains 6 chapters:

1. Irrigation and Drainage in Agriculture: A Salinity and Environmental Perspective
2. Predictive Irrigation Scheduling Modelling in Nurseries
3. Municipal Wastewater Irrigation for Rice Cultivation
4. On the Use of Decision-Support Tools for Improved Irrigation Management: AquaCrop-Based Applications
5. Technical Efficiency of the Subsurface Drainage on Agricultural Lands in the Moldova River Meadow
6. Applications of SuDS Techniques in Harvesting Stormwater for Landscape Irrigation Purposes: Issues and Considerations

All chapters are written having scholars and specialists in this field as target audiences.

Dr. Suren Kulshreshtha
University of Saskatchewan
College of Agriculture and Bioresources
Saskatoon, Canada

Dr. Amin Elshorbagy
University of Saskatchewan
College of Engineering
Saskatoon, Canada

Chapter 1

Irrigation and Drainage in Agriculture: A Salinity and Environmental Perspective

Sjoerd E.A.T.M. van der Zee, Sija F. Stofberg, Xiaomei Yang, Yu Liu, Md. Nazrul Islam and Yin Fei Hu

Additional information is available at the end of the chapter

http://dx.doi.org/10.5772/66046

Abstract

Whereas irrigation and drainage are intended to address the shortage and surplus of soil water, respectively, an important aspect to address is also the management of salinity. Plants have a limited tolerance for soil water salinity, and despite significant gaps in our practical knowledge, an impression of acceptable salinities is available for many crops. To manage soil salinity, the Leaching Requirement is an old, yet useful, concept. In this chapter, we extend this concept for soils with shallow groundwater. Particularly if shallow groundwater is saline, management is needed to avoid capillary rise of this water into the root zone. One of the tools to do so is Climate Adaptive Drainage (CAD), for which many practical gaps in knowledge remain. Also, soil mulching, of which a special case is considered in more detail, i.e., using plastic covers, may be beneficial for many purposes, including improving the water and salt balances of the root zone. However, use of plastics may have significant adverse effects. Due to water shortage, also wastewater may be re-used for irrigation. For this reason, the hazard of sodicity due to elevated Na concentrations in domestic wastewater is highlighted.

Keywords: salinity tolerance, Leaching Requirement, groundwater, adaptive drainage, soil mulching, plastic mulching, wastewater re-use, sodicity, modelling

1. Introduction

Although arid and semi-arid regions have a potential for yielding multiple crops per year, due to the large incoming radiation, this potential is profoundly limited by the small annual rainfall. An impression of the distribution of dry areas is shown in **Figure 1**, which is based on the aridity index (AI), defined as annual precipitation divided by potential evapotranspiration.

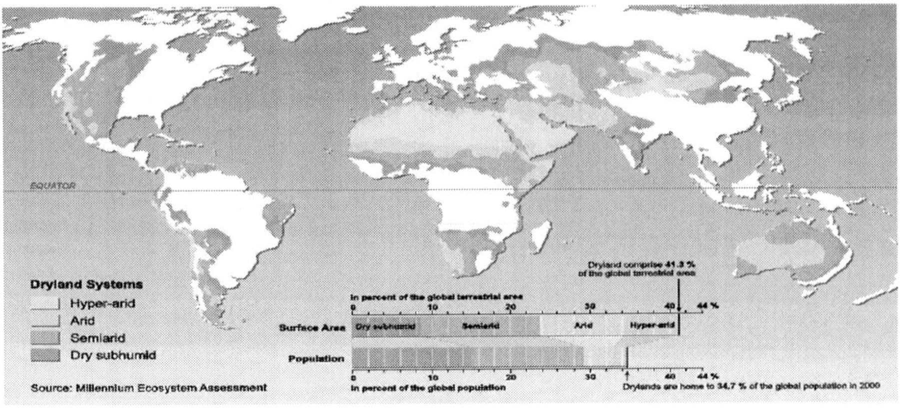

Figure 1. Distribution of dryland regions (from: Millenium Ecosystem Assessment, Chapter 22, 2005).

$$AI = P/ET_p \tag{1}$$

where P is annual rainfall and ET_p is potential evapotranspiration (both mm/y). Therefore, **Figure 1** shows the main regions, where an additional supply of water, besides rainfall, is needed or even crucial for agricultural production. However, also in the regions where water supply is sufficient, seasonal drought can be a farmer's risk. This is the case in a humid temperate country as the Netherlands.

Insufficient rain falls have to be compensated for, to ascertain sufficient agricultural yields. Irrigation using good-quality river water, groundwater and, sometimes, wastewater has been used for ages. Irrigation is often accompanied by drainage, which is the natural or artificial removal of surface and sub-surface water from a given area. Artificial drainage is always needed, except in cases with deep groundwater levels. In most river delta regions, drainage is a must to avoid water logging. This is not the only reason, because avoiding salinization is a recognized aim of drainage, as is illustrated in this chapter.

Research on irrigation and drainage has been considerable and this is not surprising as irrigated agriculture has been practiced during the past 5000 years, e.g,. in Mesopotamia. Due to long-term changes, water availability for irrigation could decline or salinity problems could develop. The underlying cause for agricultural yield depressions upon salinization is that most terrestrial plants tolerate only a limited salinity of soil water. The term salinity is usually associated with concentrations of sodium chloride (NaCl), but may be interpreted more broadly as the presence of ions in water. Rainfall usually contains few ions, but agricultural fields are often irrigated with groundwater or surface water, where this is different. The salts that may give problems in agriculture may be derived from different sources: (i) water in contact with soil material induces physical and chemical weathering, which is associated with release of ions, of which the concentrations may be increased due to evapotranspiration; (ii) groundwater may be brackish or saline due to different geohydrological causes such as sea water intrusion and the presence of old marine sediments; and (iii) use of river water with some level of salinity, originating from groundwater, sea, or industry. Salinity is often expressed as concentration (mass per volume of water), electrical conductivity (EC), or total dissolved solids (TDS).

The scope of this chapter is to address a number of environment-related aspects such as salinity and recent developments that may develop into contamination problems. Our motivation is that simple ways to predict developing environmental contamination are quite limited and sometimes not even recognized. After some background information on the vulnerability of plants for salinity, guidelines are provided for anticipating the salinity levels that may be expected in the root zone, which do not require complicated models. Some other general environmental hazards associated with irrigated agriculture are identified, as related with plastic mulching and re-use of wastewater in agriculture.

2. Salt tolerance of plant species

Efforts to limit the salinity of the root zone are, at the end, aimed at limiting adverse impacts of salinity on primary productivity of crops. In a somewhat broader sense, growth of both crops and botanic species, product quality, and other measures related to plant well-being are important in the assessment of salt-induced adverse effects.

Plants exposed to elevated salinity may experience different forms of stress. Due to the high osmotic value of saline solutions, soil water may become less available for plants to accommodate their transpiration, which directly affects primary production [1, 2] in a similar way as drought. However, it is also well known that salts (e.g. involving Na^+, Cl^-) may be toxic for plants or that toxic components such as boron (B) and selenium (Se) become better bioavailable under saline conditions. For boron, the concentrations that separate deficiency and toxicity are close together [3], making an optimum availability difficult to accomplish. In addition, induced nutrient deficiency has been well documented, e.g. for iron and nitrate [4, 5]. In many dose-effect types of exposure studies of salinity, permanent stress is considered. However, in many field situations, salt stress occurs only in limited periods or one season. Typically in temperate conditions saline stress is periodic, e.g. due to tidal or seasonal fluctuations, and is followed by salt leaching conditions. However, research on salt stress duration and plant response is quite limited.

Salt tolerance has been investigated much for agricultural crops, both in field and greenhouse conditions and, particularly, for the case where salts enter the root zone. Different plant species have different salt tolerance and strategies to deal with salinity [6, 7]. It has been well established that also different genotypes of the same species may differ according to their salt tolerance ability [8]. The accumulated experimental evidence of salt tolerance functions (MH) for crops has been summarized [2], which describe the salt concentration range in the root zone where transpiration (hence primary production) is unaffected, and in the larger salt concentrations where transpiration is reduced. Since 1977, such 'MH'-functions have been adjusted and developed for other crops or for different environmental conditions [9].

MH-functions make it is possible to compare different plant species with regard to their vulnerability to salinity.

The MH-functions describe how soil or soil water salt concentrations affect primary production. They seem to differ for different soil types and climate and weather conditions. MH-functions, apart for comparing species for salt vulnerability, are also important for predicting crop yields, because most modelling uses the MH-concept [10–13]. The aim of such modelling is often to really predict yields, e.g. if climate changes, or to generalize experimental observations to other soils or geohydrological situations. MH-functions are often used to measure water availability and salinity of the root zone throughout the growing season, which are relatively constant. However, in the field, this is generally not the case. Particularly for distributed modelling, additional assumptions on how plants spatiotemporally deal with variability are needed [14]. For instance, plant roots may compensate for dry or saline layers in soil by taking up water from other layers, but do they do so based on the water content, water potential, or a mixture of these properties? At the moment, the experimental evidence is predominantly based on water contents and salt concentrations that are as homogeneous as possible. Compensation strategies may differ according to plant species or even genotype.

In agricultural modelling, emphasis has been on salinity effects on the osmotic potential of soil water and its availability for plants. Ion specific effects as toxicity or induced deficiency are important, but generally disregarded in modelling. The reason is that crops are often well fertilized, but also that for modelling ion-specific effects, a multicomponent approach is needed as in HP2, UNSATCHEM, and SWAP-ORCHESTRA [15]. Not only is multicomponent modelling much more demanding with regard to parameterization, data, and computational efforts, but also data are needed for different plant species. This is not a trivial effort.

As much as crops have been investigated for salt tolerance, this not the case for landscape/decorative plants. Native vegetation has been investigated much from the perspective of plant ecology [16] or mapping [17]. Spatiotemporal variability of abiotic factors is often ignored, although salt tolerance of species and systems is estimated by investigating their spatial distribution and plant traits. Out of thousands of papers, only about 50 dealt with deciduous trees [18, 19]. Nevertheless, also for these categories of plants lists have been produced for their salt tolerance [20, 21]. In some cases, such listings give a quantitative grading, but a relative classification may be more appropriate [22] because the experimental techniques, quality of data, and used metrics to describe the damage may differ profoundly among papers. In view of the limited data for non-crop plants, at this moment modelling of climate change on the distribution of botanic species may be hard or impossible.

An important salt exposure pathway is that of salt spray and sprinkler irrigation. Although the importance of exposure of the leaves has been demonstrated [21, 23–25], this pathway is generally ignored in modelling. Mostly, it is mentioned that uptake of salts (notably Na and Cl) by the leaves and toxicity lead to necrosis, but osmotic effects are seldom mentioned. Timing of sprinkling plays a role, because sprinkling at night, or at frequencies that limit the duration that leaves are wet, appear to affect the degree of adverse effects: the impact is smaller as leaves are wet shorter, or transpiration is smaller. The direct (leaf exposure) pathway was also recognized [26] to enhance the impact of exposure through the root zone.

Browsing through the literature, it is clear that salt effects depend on plant species, genotype, salt composition, soil type, climate and weather conditions, and geohydrology. These factors all modify the plant's response to salinity, but some are ignored completely (ion specific effects; leaf exposure) in crop and soil models. The large complexity and significant gaps reveal that data can be fitted and described by models, but true prediction may be feasible only for broad features and large uncertainty bands.

3. Irrigation and salinity

All sources of soil water, except for artificially desalinated (deionized) water, have one thing in common: they all contain dissolved salts (together with non-ionic compounds dissolved compounds are comprehensively called solutes) to some degree. In case of rain water, little salt is dissolved but in case of brackish or saline water, salt may be present in such quantities, that the density of the liquid is increased. Thus, sea water contains 35 g/l of salt (mainly NaCl) and has a density of 1.025 g/l or 1025 kg/m^3. These density differences may have a significant effect on water flow, as will be considered later in this chapter. Groundwater and river water often contain solutes that originate from the geological formations that the water has flown through, as well as from biological or human activities. These ions include calcium (Ca^{2+}), magnesium (Mg^{2+}), carbonates (HCO_3^-), sulphates (SO_4^{2-}), nitrates (NO_3^-) and phosphates (PO_4^{3-}).

When water evaporates from the soil or is transpired by plants, most salts are left behind in the soil. Therefore, it is necessary to be alert for the hazard that too large quantities accumulate in the soil solution. This understanding was communicated already in the first half of the previous century, and described in the famous Handbook 60 [27]. For illustrative purposes, the Leaching Requirement (LR) is of use. It is based on annual water and salt balances for the root zone or 'plough layer' (here the term root zone is used). In words, both balances tell how large the (water or salt) accumulation rate is in dependence of the inflow of irrigation and rainfall water and the outflow of drained water. On the longer (multi annual) term, the water balance is zero, i.e., the net water accumulation or depletion is zero. If this were not the case, the soil would become completely water saturated, or water deficient (note: at the scale of one or several years, the water balance may be positive or negative, but on the long run, this cannot be the case as otherwise we deal with an inundated soil or completely dry one). Hence,

$$D_{iw} = D_{ET} + D_{dw} \tag{2}$$

where D stands for a 'depth of water layer', as related to a flux through a designated area (in m^3 m^{-2}y^{-1} or in mm y^{-1}), *iw* stands for irrigation water, *ET* for evapotranspiration, and *dw* for drained water. For the salt balance, it is illogical that the balance is zero: salts may accumulate or be depleted until no salt is present. Accordingly, the salt balance would be

$$\frac{\Delta\theta C z_r}{\Delta t} = D_{iw}C_{iw} - D_{ET}C_{ET} - D_{dw}C_{dw} \tag{3}$$

where θ is the volumetric water fraction, C the salt concentration, z_r the root zone thickness, and t is time.

Usually, water taken up by plants for evapotranspiration contains some salts (and nutrients), but far less than in irrigation water for those cases where salinity is an issue. Therefore, the term for ET can be neglected and Eq. (3) can be modelled for if more salt is coming in than is in equilibrium with initially present salt. The salt concentration increases, giving a behaviour as is shown in **Figure 2**.

As **Figure 2** reveals, short term variability can lead to an erratic pattern of root zone salinity. Although this variability may be important, for instance if it is a seasonal pattern, particularly the long term level at which salinity stabilizes is of main interest in a sustainability assessment: will the adopted management (involving how much is irrigated, how much is leached, in view of the quality of irrigation water) lead to unacceptable salinization or not?

On the long term, the salinity acquires the level of the dashed line in **Figure 2** and does not change anymore: the term to the left of the =– sign of Eq. (3) becomes zero. Then, Eq. (3) can be recast as follows:

$$D_{iw}C_{iw} = D_{dw}C_{dw} \tag{4}$$

where we remember that $D_{iw} = D_{ET} + D_{dw}$.

Already in Handbook 60 [27], it was realized, that the concentration in water that drains out of the root zone is equal to the concentration in the root zone, if this zone is at field capacity. Hence, $C_{dw} = C_{soil,FC}$. Inserting this in Eq. (4) and re-arranging leads to

$$\frac{D_{iw}}{D_{dw}} = \frac{C_{soil,FC}}{C_{iw}} \tag{5}$$

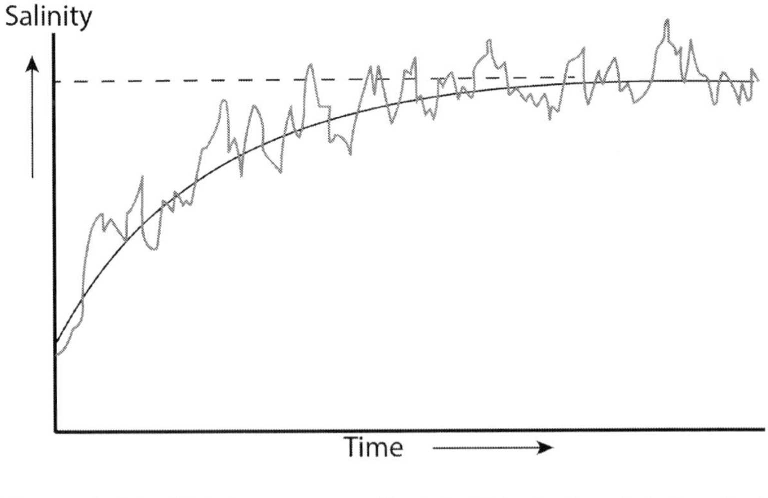

Figure 2. Increase of salinity (θCz_r) of a root zone according to Eq. (3) for short term fluctuations of water or salt concentration (grey line) or constant irrigation water concentrations and rainfall, ET and drainage rates (black line).

For illustration, this is a very useful result. Usually, the quality of irrigation water (here represented by its salt concentration) is known, as it is easily measured. Then, it is easy to determine with Eq. (5) how much irrigation water must be applied (in view of the amount of evapotranspiration, represented by $D_{ET} = D_{iw} - D_{dw}$) to drain enough to keep the soil salinity ($C_{soil,FC}$) below the level that a certain crop demands.

Practically convincing is it, to show what happens if one does not drain at all. In that case, $D_{dw} = 0$, and the division by zero implies that $C_{soil,FC}$ becomes infinitely large. The compelling message of this simple illustration is:

1. If regular leaching of salts out of the root zone does not occur, salinity will increase unbounded: concentrations continue to increase.

2. This increase of salinity may take time, particularly if irrigation water is of good quality, but it is still inevitable. On the long term, salinity must increase if salts are not leached, and agricultural soil will become too saline for agriculture, as old civilizations in Mesopotamia testify [28]!

3. Really unlimited increase of the concentration will not occur, of course, in reality. This is due to chemical precipitation of salts as soon as the solubility product is exceeded. Where this solubility product is salt as well as environment dependent, it defines an upper limit to salt concentrations in solution. The environment dependency is related with e.g. temperature, but also the solution composition. For instance, gypsum is known to be better soluble under saline (or more general: higher electrolyte concentration) than low salt conditions.

4. Impact of groundwater

To determine the need for leaching salts without consideration of groundwater is often inappropriate. In many regions, and particularly in lowlands as coastal plains and delta areas, groundwater is shallow. For the salt-affected Hortobágy area of Hungary, the guideline was developed, that if groundwater is shallower than say 3 m below the soil surface, the upward capillary flux may be significant enough to be important for the water and salt balances [29]. Simulations support this guideline [30].

The relationship of the impacts on the root zone wetness between natural precipitation (mostly rainfall) and groundwater (capillary rise) can be complex. Whereas rainfall is an independent forcing, capillary rise of water depends on the root zone water content pressure, soil hydraulic properties, and groundwater depth: root zone water affects capillary rise and is affected by it. The same is true for infiltration and runoff [31]. The complexity was one of the reasons of a new school of Ecohydrology to analyze low-dimensional flow models [11].

In the simplest form, that approach of modelling considers a root zone of which the volumetric water fraction θ depends on infiltration, evapotranspiration, and drainage. At the same time, θ is the variable that determines these fluxes (plus runoff). Vervoort and Van Der Zee [32] extended such models with capillary rise of ground water. These water balance models represent the layer between root zone and groundwater level with a functional dependency of θ and these fluxes.

The impact of seasonality on the water dynamics can be profound, and many countries are characterized with seasonality. This is also the case for root zone salinity, because dissolved salts (of alkali and earth alkali cations) respond quickly on the water fluxes. For this reason, the salt concentration of the root zone often shows a distinct short term fluctuation as in **Figure 2**. Using the ecohydrological framework with minimalist (parsimonious) models [11, 32], salt accumulation was investigated [12]. The latter took capillary rise of brackish groundwater into account. Their root zone water balance, was therefore given by

$$\varphi Z_r \frac{ds}{dt} = P - R(s) - ET(s) + U(s) - L(s) \tag{6}$$

where φ is porosity, Z_r is root zone thickness, s is water saturation (about equal to θ/φ), t is time, P is rainfall or precipitation rate, R is runoff rate, ET is evapotranspiration rate, U is capillary upward flow rate, and L is leaching or drainage rate.

The complexity of Eq. (6) is that the rainfall is erratic: rain occurs with irregular intervals and each shower has a different intensity and duration. Therefore, water saturation s and salt concentration c vary irregularly with time.

The salt mass balance is given by

$$\frac{dM}{dt} = \varphi Z_r \frac{dsC}{dt} = U(s).C_Z - L(s)C \tag{7}$$

with M the total mass of salt, C_Z is the salt concentration in the groundwater (fixed) and C is the salt concentration in the root zone. In Eq. (7), it is assumed that the other water fluxes (P, R, ET) do not transport appreciable quantities of salt, which is often reasonable. Solving Eqs. (6) and (7) shows patterns similar as **Figure 2**. If saline soil is washed with good-quality water, the mean concentration decreases but short term variability still occurs. If seasonality is profound, intra-annual periods of drought will often coincide with periods of root zone salinity, both having a tendency to reduce crop yields.

Taking into account the often erratic variability of weather [33], recently the salt accumulation in the root zone was investigated, considering that root zone salinity levels respond fast to the water quality of incoming fluxes (rainfall, irrigation, capillary rise). In general, say in 10 years, the root zone concentration attains a dynamic equilibrium with rain, irrigation water, and groundwater quality.

For water managers, it is important to be able to assess whether intended water and salt management is sustainable on the longer term: will the use of irrigation water with a known salinity lead to too much salinity? And if it does, to what degree: will crops that are planned to be part of the rotation scheme show reduced yield? If for such questions only the longer term level of the concentration in the root zone is important (i.e., the plateau in **Figure 2**), a simpler assessment may be sufficient, than analyzing Eqs. (6) and (7).

http://dx.doi.org/10.5772/66046

Comparing simulated salt concentrations using a root zone model [12], and analytical approximations as Eq. (8) that are as simple as the Leaching Requirement approach, a good agreement was obtained, as shown in **Figure 3**, with ratios close to one. Apparently, the longer term mean concentration levels can be quantified well if short term variability of weather is ignored using

$$\langle C \rangle = \frac{\langle D_{cr} \rangle C_Z + \langle D_{irr} \rangle C_{irr}}{\langle D_{dw} \rangle} \tag{8}$$

where brackets <.> refer to time averaging, subscripts *cr*, *Z*, *irr*, and *dw* stand for capillary rise, groundwater level, irrigation water, and drainage water, and *D* is the respective flux in mm or m water layer for the averaging time (usually one year, so *D* in mm/y). Eq. (8) holds if both groundwater (concentration C_Z) and irrigation water (concentration C_{irr}) carry significant salt quantities, and implies that the mean soil concentration, and the mean leaching concentration, is given by a weighted average concentration of the water fluxes.

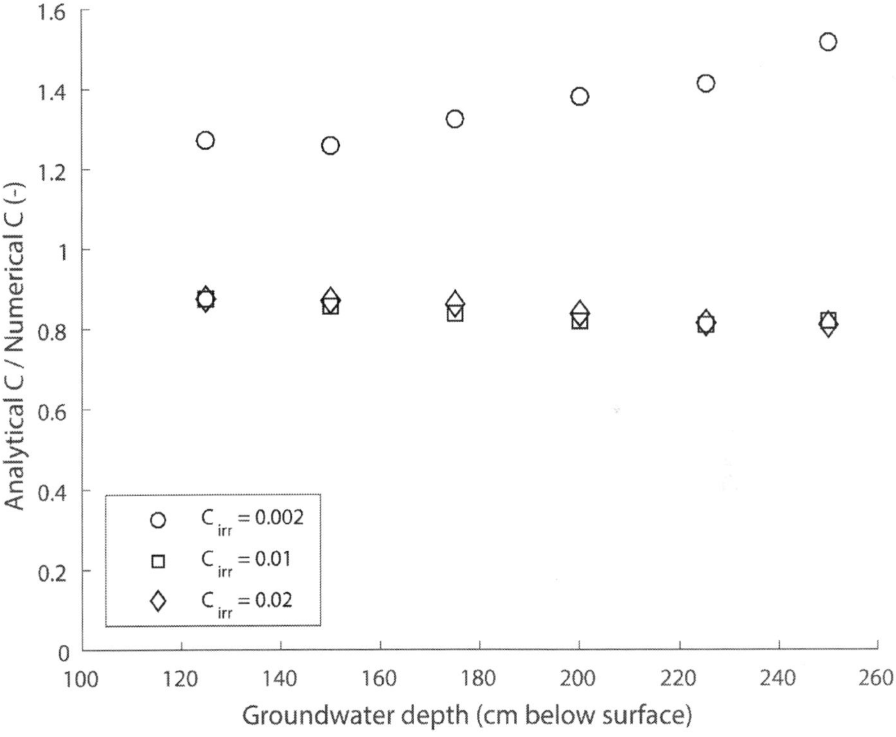

Figure 3. Ratio of calculated root zone salt concentrations with Eq. (8) and as modelled numerically, as a function of groundwater level (Z_f in cm below surface) for a temperate climate as in The Netherlands. Irrigation water salinity given by C_{irr} in mol l^{-1}, and groundwater salt concentration of 0.02 mol l^{-1}. Adapted with permission from Ref. [33].

Irrigation can be modelled similar as rainfall, except that it is a managed source of water (timing is a managing decision) and that it may contain more dissolved salts than is common for rainwater. To predict salinization using simple means, for cases not considered by the Leaching Requirement (e.g. if capillary rise is important, and possibly it contains appreciable dissolved salts, if both rain water and irrigation water infiltrate, with different salinities) is useful and Eq. (8) is an example of such a simple management tool.

With numerical models and with approximations as LR and Eq. (8), a judgement can be made whether root zone salinity is expected to become too high under the current practice, and how much and how good quality irrigation water will be needed to limit salinity to desirable levels. These tools, however, are generic and for a good desalinization of soil, the local conditions need to be known well. Besides the management decision to install a drainage system, various measures can be taken to homogenize fields (e.g. surface level) and improve the efficiency of salt leaching. Tools as discussed in this chapter may fine-tune the leached salt concentrations to agree with the farmer's needs and the salinity in discharged drainage water that are acceptable downstream.

In most model approximations for the long-term salinity of the root zone, it is assumed that level and quality of groundwater are unaffected by the root zone water and salt balance. This is a simplification. Root zone drainage water may affect groundwater, and cause fresh water bodies to develop in coastal dunes [34]. The thickness of the freshwater bodies floating on top of saline water as follows from Archimedes' law. Various approximations have been developed since the nineteenth century for the thickness of fresh water lenses, assuming a sharp interface between fresh and saline water and different assumptions on the outflow at the edge of the lens and movement of the saline water [35] (**Figure 4**).

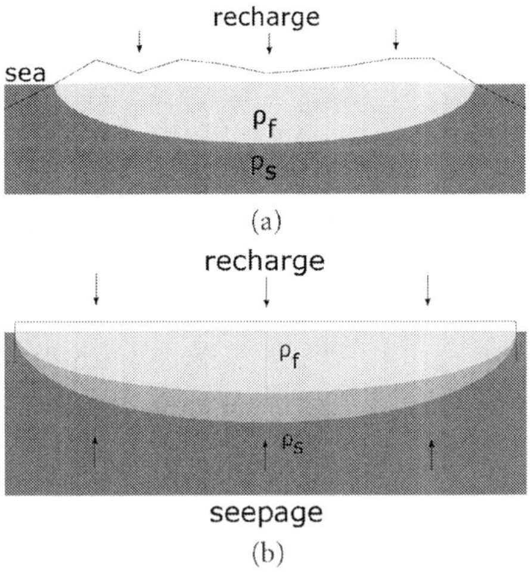

Figure 4. The freshwater lens (a) in a dune (the classical Ghyben-Herzberg lens) and (b) in an agricultural lowland field as considered by Refs. [35, 36].

Mostly, fresh water lenses outside the dune areas have been studied for situations where the soil surface is close to the sea level, and open water is at or even below this level. Then, groundwater seeps upward into the open water, and thin fresh water lenses form, that may respond quickly to changes in rainfall and evapotranspiration. They are found not only in coastal regions [36], but also in central Australia due to incidental flooding, to form an important source of fresh water to indigenous vegetation [37, 38] and 'inland' parts of PR China.

Because small fresh water lenses are vulnerable to disappear during extended periods of drought, Stofberg et al. [33] assessed how this vulnerability depends on a number of conditions. To avoid this depletion of fresh water, active management of water storage in wet periods is needed and one of the tools may be Climate Adaptive Drainage (CAD).

5. Drainage dimensioning and advanced options

Drainage of water is needed to avoid water logging, poor mechanical behaviour, and increasing salinity. This recognition has made drainage technology as well as theory to become topics of high importance in soil, water, and agricultural science. Due to the good economic situation, high population density, and usually shallow groundwater table, drainage has particularly been investigated in detail in The Netherlands. Basic theory has been developed by, e.g. Hooghoudt and others.

Dimensioning of drainage for (often flat) fields in agricultural use, is commonly aimed at management of groundwater levels to avoid water logging, poor tractability, structure degradation due to grazing cattle, and water saturated and anaerobic conditions. In their simplest form, the steady state water level between two ditches or drains can be obtained from the combination of Darcy's law and the water balance, with the Dupuit assumption. For a steady-state rainfall rate of P, we obtain

$$P = q = \frac{4K(H^2 - D^2)}{L^2} \tag{9}$$

where P is recharge (net infiltration) rate (m/day), q is the discharge into the drain or ditch (m/day), K is the hydraulic conductivity of soil (m/day), H is the water level halfway two drains measured from the impervious basis of the phreatic aquifer (m), D is the drain level measured from the same basis (m), and L is drain spacing (m). With Eq. (9), it is possible to determine the spacing needed to limit D to a value sufficiently below the soil surface.

As was already mentioned, in many climate zones it may be important to save water, for instance because precipitation may occur in a season that is unfavourable for crop production, and primary production in general. One of the tools to enable temporary storage is Climate Adaptive Drainage (CAD). In CAD, the level of drainage can be dynamically varied to improve flexibility with regard to water stored in the soil profile. The basis of this technique is quite simple: drainage, e.g. using tile drains, is no longer passive but active by

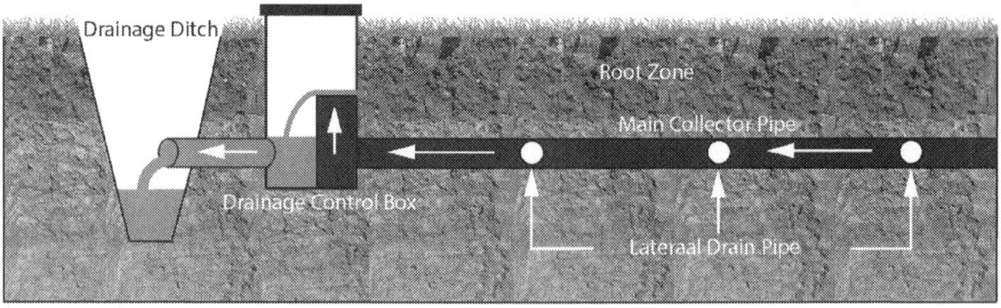

Figure 5. Climate Adaptive Drainage (CAD), with drains merging into a collector pipe, that discharges into a control box. The level in this box can be adjusted (manually or remote control), to anticipate on weather conditions in the next weeks or months.

using a collector between the drain and the receiving ditch. The collector (**Figure 5**) has an overlet shown in black of which the height can be varied. If this height above the base of the collector's overlet becomes larger, the groundwater becomes more shallow, closer to the soil surface and the root zone. Hence, adjusting the overlet height is the same as manipulating groundwater level.

CAD is beneficial to dynamically store water for the dry season. In regions with saline groundwater, and particular in areas where such saline groundwater seeps upward, as in deltas, dynamic drainage can be very useful to diminish the upward saline fluxes. With a shallow groundwater level enabled by CAD in the wet season, the upward saline seepage can be diminished. In the growing season, fresh water is consumed for evapotranspiration, and CAD can lower the saline water table, or perhaps be used to introduce better quality into soil.

6. Plastic mulching

Irrigation is often combined with other techniques that improve the water use efficiency (WUE), including several cropping and tillage techniques [39, 40]. Mulching involves putting a barrier between soil and atmosphere, using different materials such as crop residue, such as straw [41], leaves, paper, old carpet, plastic (**Figure 6**) or gravel. It may serve purposes other than water use efficiency as well, such as weed growth reduction.

In China, plastics have been used for many years after importing from Japan in 1970s. The technology has been transformed and developed and become important in agronomy, especially in NW China, where agriculture became much dependent on plastic mulching. In China, it is mainly used for cash crops, such as cotton, tobacco, medical herbs, bean nuts, loofah, and crops such as maize, bean, and potato.

Mulching serves various purposes.

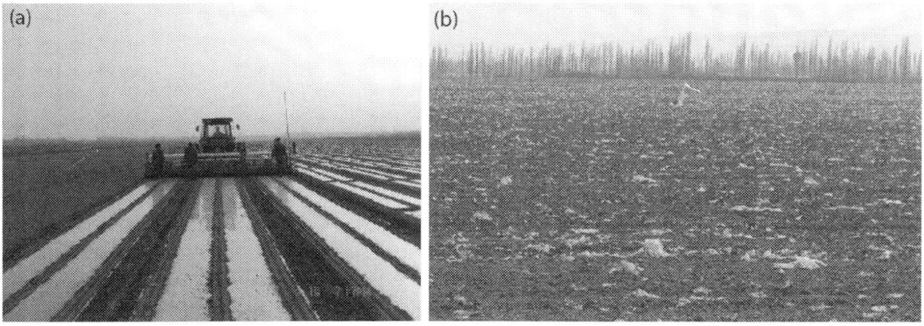

Figure 6. (a) Installation of plastic mulch and (b) postharvest field contaminated with plastics in Xinjiang, China.

1. Increasing soil temperature. Plastic mulching decreases heat exchange between soil and atmosphere, and leads to steadily increasing soil temperatures. For NW China, topsoil temperatures were found to increase 3–6°C in the top layer, and some heat is transferred into deeper layers. The sowing or planting time can therefore be earlier in spring (10–20 days) increasing the growing season. It is reported that crop yields can increase 20–30% compared with situations without plastic mulching and also crop quality can improve.

2. Reducing water evaporation and water conservation. The plastic blocks water vapor exchange between soil and open air and reduces soil evaporation. However, infiltration of rain water is not affected. Hence, soil moisture in 0–20 cm layer may increase 2% by plastic covering and in semi-arid and arid areas, around 1950 m^3 ha^{-1} water can be saved during the growing season.

3. Enhanced fertilizer efficiency. This beneficial effect is attributed to the reduction of nutrient losses via leaching, runoff and erosion. There are indications that soil available nitrogen, phosphorus and potassium might increase 30–50%, 20–30%, and 10–20%, respectively.

4. Improved solar light irradiation efficiency. Transparent plastic film and the water droplets which adhere onto this film improve soil absorption of light, benefitting photosynthetic intensity and postponing the senescence time of lower leaves of plant. It is reported that lower leaves can capture 12–14% extra catoptric light with plastic mulching, and tomato can capture 3–4 times more light, photosynthesis intensity may enhance 13.5–46.8%, and chlorophyll content in tomato may increase 5%.

5. Improving soil physical properties. Due to mulching, soil temperature rises, vapour pressure effects have been claimed to enlarge the pore sizes within soil particles, increase soil porosity, reduce soil bulk density and increase soil aggregate stability. This affects physical properties as well as soil fertilizer, water, atmosphere and heat balances, besides protecting the soil surface, avoid raindrop detachment and erosion of soil. Weed control may be less frequent and reduce soil compaction.

6. Improving soil microbial activity. By positive impacts on temperature and soil moisture, mulching is beneficial for soil microbial activity, microbial reproduction and growth, and organic matter decomposition and mineralization in the root zone, and improves nutrient availability. Also plastic use might inhibit diseases and pests.

7. Avoiding soil salinity. In saline regions, salts can be transferred by water evapotranspiration to the soil surface. By its water conserving effect, this transport is reduced.

Great changes have occurred since plastic mulching became use in China. Crop yields and quality improved. However, at the same time, plastic film residues have become a pollutant in the plough layer of agricultural soils. About 313.3×10^4 ha farmland is covered by plastic mulching, or 75% of the entire farming land of Xinjiang province, with an annual weight of plastics of 18.5×10^4 tons. In the main land for cotton planting, the plastic film residue is 25.32 g m^{-2} and has become a big problem for soil quality and crop growing. Recycling and natural degradation have therefore become important.

In China, plastic film residue recycling is mainly a human labour and machine-supported effort. About 60% of the plastic film residue is recycled with machinery and the main difficulty is the low intensity of residue and plastic pieces. To strengthen film intensity, China government has already issued a regulation since 1990s, entitled GB13735-92, for standard thicknesses, that changed from 0.008 ± 0.003 mm to 0.01–0.02 mm. However, the efficiency of machine recycling is limited. The paradox is that if the thickness increases, the cost will also be increased which farmers cannot afford. But if plastic film residue is recycled manually, the cost will become higher, especially in the regions with a labour shortage and efficiency of collecting plastic is low. Hence, machine supported recycling needs further development.

One of the concerns with all this use of plastics is that they may be gradually transformed into microplastics: particles smaller than 5 mm. Particularly for marine ecosystems, microplastics and their effects on various organisms have been investigated [42]. Also the plastics may interact with other pollutants accumulation in organisms [43], and possibly affect their impacts. For soil organisms, the study of microplastics is still in its initial stages as far as accumulation and effects is concerned, and one of the few papers addressing this issue is very recent [44]. Therefore, it is quite well possible, that bio-availability of pollutants and of microplastics, and their adverse environmentally and ecological effects may be linked, and this deserves closer investigation.

An alternative to deal with plastic residues in soil is to use biodegradable plastic film. However, such types of plastic film tends to be unstable for the range of natural conditions encountered, and therefore, these plastic films have not been applied widely yet. The use of organic mulch, such as crop residues, may be more sustainable from a chemical and biological perspective, as it would not lead to pollution. At the same time, it may have benefits regarding organic matter content, nutrient recycling and soil biology [45]. Although many studies are available regarding the effect of plastic cover on WUE, there are few studies available that compare the use of organic mulch to plastic covering. The results of these studies do not seem

http://dx.doi.org/10.5772/66046

very promising for using organic mulches as an alternative for plastic, with lower water use efficiencies [46]. The use of plastic mulch was compared with the use of traditional practice and crop residue mulching [47]. It appeared that using plastic mulch not only led to higher yields in two of the three studied years, but also lead to significant soil water depletion in deeper soil layers, which may be explained by increased evapotranspiration [48]. However, in warmer climates, straw seems to conserve more soil water than plastic cover, although the effects on yield may differ [49, 50]. One study not only reported positive effects of straw mulching on salinity, but also mentioned that many uncertainties still remain [51].

7. Re-use of wastewater and soil sodicity

To make optimal use of resources, also marginal water, such as sewage or domestic wastewater is used for irrigation, in rural and peri-urban agriculture. However, this 'wastewater' suffers often of poor quality.

Compounds in wastewater that may be beneficial are nutrient elements, such as C, N, P, and K, that make wastewater irrigation similar to fertigation. Nutrient supply via wastewater is generally unbalanced, compared with the needs of crops. This may lead to accumulation in soil, or excessive leaching to groundwater and surface water.

Several other elements that may cause adverse (toxic) effects, for plants, soil fauna, animals and humans are boron, selenium, and (heavy) metals. It is worthwhile to mention the presence of Na in most domestic wastewaters. Sodium is often present due to the high salt concentration in the human diet. The hazard of Na-fluxes into agricultural root zones irrigated with wastewater cannot be judged well by only measuring the aqueous Na-concentration. As all cations, Na may sorb onto suspended materials such as organic matter (or organic carbon) and the load of Na that reaches soil with organic carriers might be appreciable. Moreover, also other cations (Ca, Mg) sorb, preferentially, onto suspended and dissolved (DOC) organic carbon. This may make the effect of Na larger, by suppressing the beneficial effect of Ca and Mg. Hence, the adverse effects of adding Na via domestic wastewater needs to be anticipated on the concentrations of Na, Ca, and Mg, as well as Biological Oxygen Demand (BOD), as a measure of suspended and dissolved organic matter.

The adverse effect of sodium is related to it being monovalent: Na^{+}. Accordingly, it may react with mineral and organic surfaces and surface groups, but different from multivalent cations as Ca^{2+}, Mg^{2+}, Fe^{3+} and Al^{3+}, it is less equipped to form charged bridges between different reactive surface groups. Therefore, why cementing agents as organic matter are known to detach from the sodic soil's solid phase and disperse and accumulate at the soil surface.

For similar reasons, sodic soils may be susceptible to structure deterioration. Being monovalent, Na-ions are less able to electrostatically neutralize the solid surface negative charge of soil clay colloids. Under saline conditions, this may not be so much of a problem, in view of the high concentration of ions that neutralize the electrostatic field of the negative clay colloids. If the salinity drops, diffuse double layers at the clay colloid surface expand (say an order of magnitude, of e.g. 40–400 nm) due to enormous swelling pressures [3]. Swelling causes very small pores (tens of nm) to swell at the expense of macropores, which are important for conveying water and (soil) air. Soil is then dominated by small pore sizes, and becomes almost impermeable to water. An impression of the reduction of the hydraulic conductivity due to decreasing salinity of a sodic soil is given in **Figure 7** [30]. In **Figure 7**, the sodicity of soil is expressed with the common metric of Exchangeable Sodium Percentage (ESP): the percentage of the cation exchange capacity that is occupied by sodium. Usually, a soil is called sodic if ESP exceeds 15%. As shown, the hydraulic conductivity decreases rapidly in case salinity decreases. In reality, the effect of sodicity on hydraulic properties is even more complex, as also the retention function is significantly affected.

Sodic soils can be found in the Central Plain of Hungary, in the Hortobágy region, developed after the regulation of the main rivers, one and a half century ago. These physically degraded soils, with distinct B_2t horizons at say 20–40 cm depth that virtually impede vertical water transport, are perhaps only useful for extensive agriculture. Due to the small thickness of the

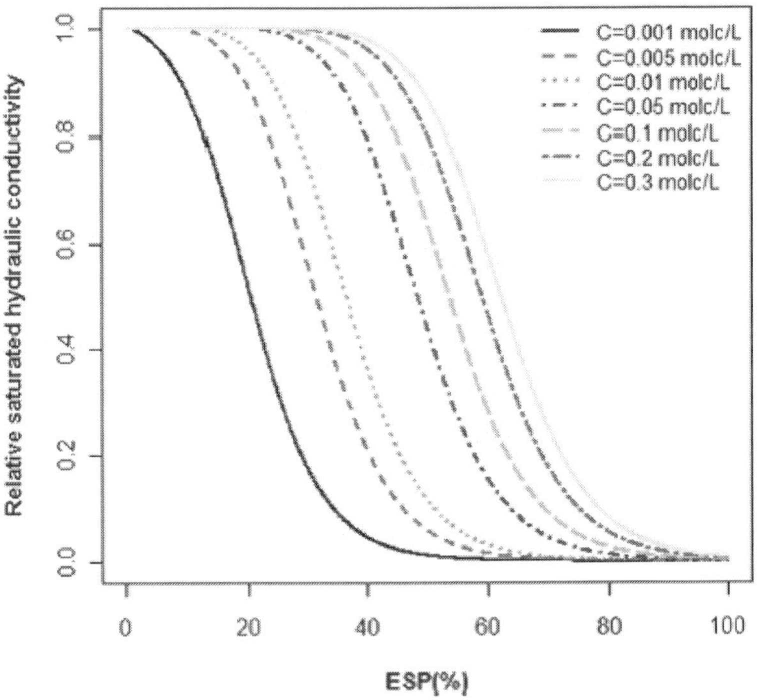

Figure 7. Relative saturated hydraulic conductivity as a function of ESP for different total salt concentrations in the soil solution (after Ref. [30]).

A-horizon, that supplies water for the crops (capillary rise from the ground water being impossible due to the B_2t horizon), the water supply is very small. Remediation during the past century proved quite challenging and a large part of the Hungarian central plain is still suffering from small yields. This indicates that sodicity is a process to avoid.

8. Conclusions

Water scarcity can be countered by improving water use efficiency, but as inevitably, irrigation leads to salinity under conditions of scarce fresh water, salts should be leached and removed. The Leaching Requirement (LR) shows how to irrigate and avoid salinity, for deep groundwater. Also for shallow groundwater, minimalist modelling as with LR may lead to very useful approximations for the estimation of the long term salinity of the root zone.

In many agriculturally important regions worldwide, such as deltas, groundwater is sufficiently shallow to be taken into account in assessments of salinization risks. Accordingly, drainage and irrigation dimensioning necessitate the consideration of groundwater levels, which are strongly related with capillary rise replenishment of root zone water. Simple guidelines, as proposed in this chapter, may become very relevant to avoid salinity. Furthermore, advanced dynamic approaches to drainage, such as Climate Adaptive Drainage (CAD), may be very useful for saving water till the dry season, suppressing saline upward seepage and capillary rise, and removing salts from the root zone more effectively. In the Dutch WaterNexus program (with the lead author), CAD is experimentally and theoretically investigated for this purpose.

Besides salinity, plastic mulching may have environmental risks in view of scarcely known impacts on environment such as soil ecology, and ultimately, human health. Likewise, sodicity, which is often a salinity-related issue, requires urgent attention: it is a stealthily developing condition, that cannot well be observed by the land user without advanced chemical tools, its adverse effects may occur quite suddenly, and remediation may require enormous efforts and significant resources. For these reasons, sodicity needs to be avoided as much and as early as possible. That requires expertise, excellent communication, and great awareness.

This chapter has been aimed at providing awareness and illustrating the value of theoretical understanding (e.g. the Leaching Requirement and further developments based on it). Such understanding is important to anticipate when currently common practice may be in conflict with long term sustainability.

Acknowledgements

This contribution was made possible through the project WaterNexus that is funded by the Dutch Technology Foundation STW and the Dutch Ministry of Infrastructure and Environment.

Author details

Sjoerd E.A.T.M. van der Zee[1,3]*, Sija F. Stofberg[1], Xiaomei Yang[1], Yu Liu[2], Md. Nazrul Islam[4] and Yin Fei Hu[2]

*Address all correspondence to: Sjoerd.vanderZee@WUR.NL

1 Soil Physics and Land Management, Environmental Sciences Group, Wageningen University, Wageningen, Netherlands

2 Xinjiang Academy of Agriculture and Reclamation Science, Shihezi, China

3 School of Chemistry, Monash University, Melbourne, Australia

4 Soil Science Division, Bangladesh Rice Research Institute, Gazipur, Bangladesh

References

[1] De Wit CT. Transpiration and crop yields. Versl. Landbouwkd. Onderz. 64, 88pp. 1958.

[2] Maas EV, Hoffman GJ. Crop salt tolerance—current assessment. ASCE Journal of the Irrigation and Drainage Division. 103 (2), 115–134,1977.

[3] Bresler, E, MacNeal BL, Carter DL. Saline and Sodic Soils: Principles-Dynamics-Modeling. Springer-Verlag, New York. 236 pp., 1982.

[4] Schinas S, Rowell DL. Lime-induced chlorosis. Journal of Soil Science. 28, 351–368, 1977.

[5] Grattan SR, Grieve CM. Mineral element acquisition and growth response of plants grown in saline environments. Agriculture, Ecosystems and Environment. 38, 275–300, 1992.

[6] Parida AK, Das AB. Salt tolerance and salinity effects on plants: a review. Ecotoxicology and Environmental Safety. 60, 324–349, 2005, doi:10.1016/j.ecoenv.2004.06.010.

[7] Munns R, Tester M. Mechanisms of salinity tolerance. Annual Review of Plant Biology. 59, 651–681, 2008, doi:10.1146/annurev.arplant.59.032607.092911.

[8] FLowers TJ, Gaur PM, Gowda CLL, Krishnamurthy L, Samineni S, Siddique KHM, Turner NC, Vadez V, Varshney RK, Colmer TD. Salt sensitivity in chickpea. Plant, Cell & Environment. 33, 490–509, 2010, doi:10.1111/j.1365-3040.2009.02051.x.

[9] Shannon MC, Grieve CM. Tolerance of vegetable crops to salinity. Scientia Horticulturae. 78, 5–38, 1999.

[10] Homaee M, Feddes RA, Dirksen C. A macroscopic water extraction model for nonuniform transient salinity and water stress. Soil Science Society of America Journal. 66, 1764–1772, 2002.

[11] Rodriguez-Iturbe I, Porporato A. Ecohydrology of Water-controlled Ecosystems: Soil Moisture and Plant Dynamics. Cambridge University Press, Cambridge, UK, 2004.

[12] Shah SHH, Vervoort RW,Suweis S, Guswa AJ, Rinaldo A, Van Der Zee SEATM.Stochastic modeling of salt accumulation in the root zone due to capillary flux from brackish groundwater. Water Resources Research. 47, W09506, 2010, doi: 10.1029/2010WR009790.

[13] Stöckle CO, Donatelli M, Nelson R. CropSyst, a cropping systems simulation model. European Journal of Agronomy. 18(3–4), 289–307, 2003, doi:http://dx.doi.org/10.1016/S1161-0301(02)00109-0

[14] Kuhlmann A, Neuweiler I, Van Der Zee SEATM, Helmig R. Influence of soil structure and root water uptake strategy on unsaturated flow in heterogeneous media. Water Resources Research. 48, W02534, 2012.

[15] Meeussen J. ORCHESTRA: an object-oriented framework for implementing chemical equilibrium models. Environmental Science and Technology. 37(6), 1175–1182, 2003.

[16] Niu G, Cabrera RI. Growth and physiological responses of landscape plants to saline water irrigation: a review. HortScience. 45(11), 1605-1609, 2010.

[17] Tóth T, Csillag F, Biehl LL, Michéli EE. Characterization of semivegetated salt-affected soils by means of field remote sensing. Remote Sensing of the Environment. 37, 167-180, 1991.

[18] Francois LE, Maas EV. Plant Response to Salinity: An Indexed Bibliography. USDA-ARS, ARM-W-6, Washington, D.C., 1978.

[19] Francois LE, Maas EV. Plant Response to Salinity: A Supplement to An Indexed Bibliography. USDA, ARS, ARS-24, Washington, D.C., 1985.

[20] Ehlig CF, Bernstein L. Foliar absorption of sodium and chloride as a factor in sprinkler irrigation. Journal of the American Society for Horticultural Science. 74, 661-670, 1959.

[21] Bernstein L, Francois LE. Effects of frequency of sprinkling with saline waters compared with daily drip irrigation. Agronomy Journal. 67, 185-190, 1975.

[22] Hop MECM. Salt tolerance of plants. Dendroflora. 47, 43-73, 2010 (in Dutch).

[23] Benes SE, Aragues R, Grattan SR, Austin RB, Foliar and root absorption of Na and Cl in maize and barley: implications for salt tolerance screening and the use of saline sprinkler irrigation. Plant Soil. 180, 75-86, 1996.

[24] Maas, EV, Grattan SR,Ogata G. Foliar salt accumulation and injury in crops sprinkled with saline water. Irrigation Science. 3, 157-168, 1982.

[25] Busch CD, Turner Jr F. Sprinkler irrigation with high salt-content water. Transactions of the ASAE. 10, 494-496, 1976.

[26] Jordan LA, Devitt DA, Morris RL,Neuman DS. Foliar damage to ornamental trees sprinkler-irrigated with reuse water. Irrigation Science. 21, 17-25, 2001, DOI:10.1007/s00271-001-0050-y.

[27] Richards LA, Allison LE,Bernstein L, Brown CA, Fireman JW, Hatcher M, Hayward JT, Pearson HE, Reeve GA, Wilcox RC. Diagnosis and improvement of saline and alkali soils. Agric. Handbook. vol. 60, 160 pp, USDA, Washington, D.C., 1954.

[28] El-Ashry MT, Van Schilfgaarde J, Schiffman S. Salinity pollution from irrigated agriculture. Journal of Soil Water Conservation. 40(1), 48-52, 1985.

[29] Szabolcs I. Salt Affected Soils. 274 pp., CRC Press, Boca Raton, FL, 1989.

[30] Van Der Zee SEATM, Shah SHH, Vervoort RW. Root zone salinity and sodicity under seasonal rainfall due to feedback of decreasing hydraulic conductivity. Water Resources Research. 9431-9446, 2014, doi:10.1002/2013WR015208.

[31] Appels WM, Bogaart PW, van der Zee SEATM. Surface runoff in flat terrain: How field topography and runoff generating processes control hydrological connectivity. Journal of Hydrology. 534, 493-504, 2016.

[32] Vervoort RW, Van Der Zee SEATM. Simulating the effect of capillary flux on the soil water balance in a stochastic ecohydrological framework. Water Resources Research. 44, W08425, 2008, DOI:10.1029/2008WR006889.

[33] Stofberg SF, Oude Essink GHP, Pauw PS, De Louw PGB, Leijnse A, Van Der Zee SEATM. Fresh water lens persistence and root zone salinization hazard under temperate climate. Water Resources Management. 2016, doi:10.1007/s11269-016-1315-9.

[34] Raats PAC. Salinity management in the coastal region of the Netherlands: a historical perspective. Agricultural Water Management. 157, 12-30, 2014, doi:10.1016/j.agwat.2014.08.022.

[35] Maas K. Influence of climate change and sea level rise on a Ghijben Herzberg Lens. Journal of Hydrology. 347, 223–228, 2007.

[36] Eeman S, Leijnse A, Raats PAC, Van Der Zee SEATM. Analysis of the thickness of a fresh water lens and of the transition zone between this lens and upwelling saline water. Advances in Water Resources. 34(2), 291–302, 2011, http://dx.doi.org/10.1016/j.advwatres.2010.12.001.

[37] Cendón DI, Larsen JR, Jones BG, Nanson GC, Rickleman D, Hankin SI, Pueyo JJ, Maroulis J.Freshwater recharge into a shallow saline groundwater system, Couper Creek floodplain, Queensland, Australia. Journal of Hydrology. 392, 150-163, 2010.

[38] Cartwright I, Weaver TR, Simmons CT, Fifield LK, Lawrence CR, Chisari R, Varley S. Physical hydrogeology and environmental isotopes to constrain the age, origins and stability of a low-salinity groundwater lens formed by periodic river recharge: Murray Basin, Australia. Journal of Hydrology. 380, 203-221, 2010.

[39] Hatfield JL, Sauer TJ, Prueger JH. Managing soils to achieve greater water use efficiency: a review managing soils to achieve greater water use efficiency: a review. Agronomy Journal. 93, 271–280, 2001.

[40] Howell TA. 561bb7ef08ae6d17308b03eb.tif. Enhancing water use efficiency in irrigated agriculture. Agronomy Journal. 93, 281–289, 2001.

[41] Huang Y, Chen L, Fu B, Huang Z, Gong J. The wheat yields and water-use efficiency in the Loess Plateau: straw mulch and irrigation effects. Agricultural Water Management. 72, 209–222, 2005, doi:10.1016/j.agwat.2004.09.012.

[42] Wright SL, Rowe D, Thompson RC, Galloway TS. Microplastic ingestion decreases energy reserves in marine worms. Current Biology. 23(23), R1031–R1033, 2013.

[43] Koelmans AA, Besseling E, Wegner A, Foekema EM. Plastic as a carrier of POPs to aquatic organisms. A model analysis. Environmental Science and Technology. 47 (14), 7812–7820, 2013.

[44] Huerta Lwanga E, Gertsen H, Gooren H, Peters P, Salanki T, Van Der Ploeg M, Besseling E, Koelmans AA, Geissen V. Microplastics in the terrestrial ecosystem: implications for Lumbricusterrestris (Oligochaeta, Lumbricidae). Environmental Science and Technology. 50, 2685–2691, 2016, DOI: 10.1021/acs.est.5b05478.

[45] Hobbs P, Sayre K, Gupta R. The role of conservation agriculture in sustainable agriculture. Philosophical Transactions of the Royal Society B: Biological Sciences. 363, 543–555, 2008, doi:10.1098/rstb.2007.2169.

[46] Li SX, Wang ZH, Li SQ, Gao YJ, Tian XH. Effect of plastic sheet mulch, wheat straw mulch, and maize growth on water loss by evaporation in dryland areas of China. Agricultural Water Management. 116, 39–49, 2013, doi:10.1016/j.agwat.2012.10.004.

[47] Zhang S, Li P, Yang X, Wang Z, Chen X. Effects of tillage and plastic mulch on soil water, growth and yield of spring-sown maize. Soil and Tillage Research. 112, 92–97, 2011, doi:10.1016/j.still.2010.11.006.

[48] Li S, Kang S, Li F, Zhang L. Evapotranspiration and crop coefficient of spring maize with plastic mulch using eddy covariance in northwest China. Agricultural Water Management. 95, 1214–1222, 2008, doi:10.1016/j.agwat.2008.04.014.

[49] Ghosh PK, Dayal D, Bandyopadhyay KK, Mohanty M. Evaluation of straw and polythene mulch for enhancing productivity of irrigated summer groundnut. Field Crops Research. 99, 76–86, 2006, doi:10.1016/j.fcr.2006.03.004.

[50] Yang Y, Liu X, Li W, Li C. Effect of different mulch materials on winter wheat production in desalinized soil in Heilonggang region of North China. Journal of Zhejiang University Science. B7, 858–867, 2006, doi:10.1631/jzus.2006.B0858.

[51] Pang HC, Li YY, Yang JS, Liang YS. Effect of brackish water irrigation and straw mulching on soil salinity and crop yields under monsoonal climatic conditions. Agricultural Water Management. 97, 1971–1977, 2010, doi:10.1016/j.agwat.2009.08.020.

Chapter 2

Predictive Irrigation Scheduling Modeling in Nurseries

Rousseau Tawegoum, Florian Leroy,
Gérard Sintes and Gérard Chassériaux

Additional information is available at the end of the chapter

http://dx.doi.org/10.5772/67341

Abstract

Water requirement allocation plays an important role in modern farming management. Evapotranspiration-based irrigation controllers can ideally provide irrigation according to the water requirements of the plant. This chapter describes predictive irrigation scheduling in nurseries with multiple crop species and high-frequency water requirements under limited resources. Based on historical data, time-series analysis is used to forecast evapotranspiration, an essential element in water balance equation. An algorithm based on a hierarchical research including dispatching priority rules and taking into account crop characteristics, available water, and constraints of the hydraulic network is proposed to predict irrigation schedules, with the objective of minimizing crop's water stress periods and optimizing resource materials. Simulation results with different climatic conditions show on the one hand the ability of the time-series model to forecast potential evapotranspiration, and on the other hand that, given a typical nursery, the proposed predictive approach of irrigation scheduling compared to the non-predictive approach makes it possible to prevent crop's water stress.

Keywords: multiple crops, high-frequency irrigation, multiobjective optimization, evapotranspiration forecasting, time series

1. Introduction

Accurate scheduling of irrigation is essential for maximizing crop production while conserving water and ensuring irrigation systems that are environmentally and economically sustainable. Effective scheduling requires good knowledge of crop tolerance to stress, crop water demand, and soil water characteristics. Water availability is one of the most critical factors in the determination of plant survival. The cost of water today represents a relatively large percentage of overall production costs. Moreover, environmental policies now tend to limit excessive use of water [1]. Various methods and tools for irrigation scheduling listed below have been developed, ranging from those based on the water status of the soil or plant, to those that use a model to estimate soil water balance. Irrigation control has been carried out using different time scales and different approaches.

In Ref. [2], a time-threshold model is defined as the expected daily amount of time that a crop exceeds its temperature threshold when humidity does not limit cooling of the canopy through transpiration, to detect a water-stressed condition and to signal the need for irrigation. In Ref. [3], a distributed irrigation control system with autonomous wireless controller units used soil water potential measurements to control the amount of water applied to each specific area of a field, and measurements of the system's hydraulic pressure to communicate together. A comparison of the performance of an irrigation-control tray method (extensively used in greenhouses for plants grown in bags for horticultural production) adapted to the specific conditions of plants grown outdoors in containers, to the tensiometric method was carried out in Ref. [4]. In both methods, irrigation was stopped after a fixed time (2–4 min, depending on the growth phase), and new irrigation cycles were not initiated for the next 120 min. This study concluded that the plant biomass was not significantly different between the two methods and that the irrigation events were comparable. In Ref. [5], a prototype of a real-time smart sensor for irrigation scheduling in cotton crop of four different management zones on weekly basis is developed. In the same area, [6] evaluated the precision of soil moisture sensors for irrigation control to measure the volumetric soil water content. The study conducted in Ref. [7] concluded that continuous monitoring measurements of trunk diameter fluctuation (TDF) could be used for irrigation scheduling in young olives tree under intensive production. TDF measurements under variable water regimes have now been published for some fruit tree species (apple and peach, [8] and mature peach trees [9]).

Authors in Ref. [10] were the first ones to develop an irrigation-scheduling program using meteorological data to calculate water use; many variations of this approach are now in use. Ref. [11] presented a study of a daily forecasting system of irrigation water requirements that can provide management support for administrators in terms of water supply and water distribution in irrigation schemes. The system is formulated by using the fuzzy theory based on analysis of water management logic that is based on the administrator's experience and knowledge. Using the results of field tests, they showed that the strategy for forecasting primarily depends on the intuitional and the creative judgment of administrators. In Ref. [12], a simple spreadsheet model uses a water-budgeting approach to schedule irrigation of a eucalyptus plantation. Their model estimated plantation water use on the basis of a pan coefficient (the ratio of water use to pan evaporation), and measured pan evaporation using the Penman-Monteith (P-M) equation. Their model calculated daily changes in soil water and salinity level by tracking various components of the water balance, and enabled the user to design an irrigation schedule by predicting future irrigation requirements based on the current rate of water use. To enable model users to estimate water use without the need for detailed climate data and complex evapotranspiration models, monthly pan coefficients were derived for 33 reference sites within 10 biogeoclimatic zones across Australia. In Ref. [13], irrigation scheduling with an automated evaporation pan system is performed using automated measurements of evapotranspiration (ET) from a screened pan. The crop evapotranspiration is accumulated hourly in a residual. The residual is compared hourly to the user-set irrigation threshold before irrigation is called for. In Ref. [14], the ability of three brands of ET-based irrigation controllers for irrigation scheduling in a standard residential landscape is analyzed in comparison with a theoretical model of soil water balance. They used a daily soil water balance model to estimate the theoretical irrigation needs and compare the latter with the actual amounts of water applied. Other irrigation methods are based upon empirical approaches, usually derived from the visual aspect of the plant and sometimes using sensors and time controllers to monitor watering. No prediction of crop water requirements is taken into account in most of these approaches. The control unit usually reacts to changes and perturbations in the environmental parameters of the nursery area.

Sustainable irrigation aims to match water availability and water needs in quantity and quality, in space and time, at reasonable costs and with acceptable environmental incomes [15]. Triggering in advance irrigations in horticultural nurseries with limited material resources could improve the production quality. In this way, water requirement forecasting could be helpful. Thus, predicting the evolution of crop water needs in a nursery area is essential to maintain the production under control and to ensure crop safety. This aspect is more particularly cogent in horticulture where short-term variations of local weather conditions may modify evapotranspiration. This variation has to be anticipated to prevent water stress. In a similar way, [16] proposed a hybrid approach combining a simplified crop transpiration model to predict the necessary water supply and water flow measurements from the crops. This approach was used to iteratively adapt the model coefficients. A crop transpiration model was then used to predict water supply and water flow measurements, while a simple model was used to adapt the model coefficients.

The diversity of container nursery production is different from any other faced of agriculture. The sizes and shapes of plants and containers, number of plant species and cultivars, methods of irrigation delivery, fertilizer types and application methods, and the number of plants per unit area make container production a very complex problem [17]. Substrate containers have a low storage capacity, crops have a high water requirement during sunny periods, and water contents fluctuate rapidly; that is why very specific watering methods should be applied to irrigate containers in horticultural nurseries. Continuous dripping systems are currently on the rise. Because evaluations of the water content of the substrate once a day are not sufficient, daily weather data afford the opportunity to react quickly to weather changes. Real-time weather data can provide information based on recent potential or reference evapotranspiration (ET_0) rates. Moreover, irrigation networks usually supply only a limited number of plots because the main line has a limited capacity; this makes irrigation control complex for nursery workers. Water availability is the most crucial factor for plant survival and development. Consequently, water requirement forecasts are valuable tool for irrigation management.

The primary objective of irrigation decision making is to apply enough water so plant growth is not restricted. Minimizing leaching by monitoring container drainage and adjusting irrigation scheduling accordingly is not the primary objective for most growers at this time. Few growers are using BMPs (best management practices) such as ET-based irrigation scheduling, tensiometers, or other systems of objective irrigation scheduling [17]. Simulation models would use local weather to help growers with BMP decision making, including irrigation and nutrient scheduling.

In a former study [18], an example of irrigation triggering based on ET_0 prediction was presented. Results with two experimental plots showed noticeably that most of the irrigation events of the predictive triggering took place earlier than in the nonpredictive triggering. This chapter focuses on large-scale multiple crops nurseries, with high-frequency irrigation requirements under limited water availability, with the objective of scheduling simultaneous irrigation requests of the crops. The chapter is divided into two parts. The first part is devoted to ET_0 forecasting. It first describes the SARIMA (Seasonal AutoRegressive Integrated Moving Average) structure model used in time-series analysis, and then presents identification of the model coefficients and the validation results. The second part focuses on the predictive irrigation scheduling. The main structure of the scheduling algorithm is then presented and is followed by the comparison of simulation results between the nonpredictive irrigation-scheduling and the predictive irrigation-scheduling approaches.

2. ET_0 prediction

Evapotranspiration is defined as the evaporation from a soil surface and the transpiration from plant material [19]. Reference ET is described as the ET from a hypothetical reference crop with the features of an actively growing, well-watered, dense green cool season grass of uniform height. Many equations are available in the literature for ET_0 estimation. The most precise one accepted by the international scientific community is the Penman-Monteith (P-M) equation for its good results compared with other equations in various regions worldwide [19].

The FAO-56 PM equation for the hourly time step reads as follows:

$$ET_0 = \frac{0.408\Delta(Rn - G) + \gamma\left(\frac{37}{T_{hr} + 273}\right)u_2\left(e^0(T_{hr}) - e_a\right)}{\Delta + \gamma\left(1 + 0.34\,u_2\right)} \quad (1)$$

where e_a is the actual average hourly vapor pressure (kPa), $e^0(T_{hr})$ the saturation vapor pressure at T_{hr} (kPa), U_2 the average hourly wind speed (m s^{-1}), T_{hr} the mean hourly air temperature (°C), γ the psychrometric constant (kPa °C^{-1}), G the soil heat flux density (MJ m^{-2} h^{-1}), Rn the net radiation at the grass surface (MJ m^{-2} h^{-1}), Δ the slope of the saturation vapor pressure curve at T_{hr} (kPa °C^{-1}), and ET_0 is the reference evapotranspiration (mm hr^{-1});

ET_0 gives a potential evapotranspiration value issued from modeling. The relationship between that derived value and the exact amount of water required by the plant depends on the crop coefficient, that is, a biometric parameter that varies with the crop species and its growth stage (height of the aerial part).

2.1. Model structure

Using reference evapotranspiration ET_0 as an indicator for triggering irrigation can offer an advantage for nursery workers since mixed-farming irrigation control is hard to tackle. A few studies deal with evapotranspiration forecasting using time-series analysis. In Ref. [20], time-series modeling was investigated to forecast the monthly reference for crop evapotranspiration. Paper [21] proposed a daily irrigation-scheduling algorithm based on ET prediction. An

ARIMA (AutoRegressive Integrated Moving Average) model was also used in Ref. [22] to forecast daily and hourly references for evapotranspiration. In the latter study, the analysis evidenced a wide scattering of calculated versus forecast values, especially for hourly values.

The accurate forecasting of ET_0 in nurseries based on prevailing meteorological conditions could lead to an efficient management of plot-valve opening. Although meteorological centers have a huge computational power, the weather forecasts used to calculate ET_0 are only accurate on a regional scale, with lower performance at the local scale. Such poor performance can be explained by the fact that current weather forecasting uses 10-km wide coarse elementary square meshes. Thanks to the recent developments in supercomputers and observing systems, the results from the latest research in numerical prediction of weather systems achieve meshes between 4 and 2.5 km wide in the national weather services of a limited number of European countries [23].

In order to compute hourly ET_0, a climatic database with four types of measurements (global radiation, air temperature, relative humidity, and wind speed) was used. The meteorological parameters were made available every 10 min. At the end of each hour, a computation of ET_0 was obtained by averaging these six measurements. As the wind velocity is one of the most difficult parameters to forecast accurately, the reliability of the forecast using physically based equation was reduced. Thus establishing a separate model for each component of the climatic data would have led to uncertainties in the final ET_0 value.

Like the four climatic data elements it is based on, ET_0 can be considered as a time series because it corresponds to a set of *N* successive random observations $x_1, x_2, \ldots, x_N$, performed at a specific frequency. ET_0 can also be regarded as a specific outcome of a statistical process. Most time series are stochastic in that the future is only partly determined by past values; as a result, it is impossible to reach exact predictions: they have to be replaced by the notion that the probability distribution of future values is determined by past values.

The original time series had a 24-h periodicity that corresponded to the normal evolution of the meteorological parameters, more particularly radiation that plays determining role in ET_0 estimation. This periodicity allows for well-adapted SARIMA models as compared to the AR (Auto Regressive), MA (Moving Average), and ARMA models we also tested. In these models, results are not accurate enough, and there are too many parameters to be estimated. The SARIMA model integrates seasonal fluctuations; we used the estimation procedure suggested by Ref. [24] in this context.

The general multiplicative SARIMA model of order $(p,d,q) \times (P,D,Q)_S$ is defined as follows:

$$\phi_p(B)\,\Phi_P(B^s)\,\nabla^d\,\nabla_S^D\,Z_t = \theta_q\,\Theta_Q(B^S)\,a_t \tag{2}$$

where

$$\nabla = 1 - B \tag{3}$$

$$\nabla_s = 1 - B^s \tag{4}$$

∇ is the differencing operator, ∇_s is the seasonal differencing, *B* is the backward shift operator. a_t is a purely random process (corresponding to a zero-mean Gaussian white noise with variance σ_ε^2), Z_t and are formed from the sampled original series at time *t*, *S* is the period of the series, *d* is the ordinary differencing order, and *D* is the seasonal differencing order. ϕ_p, Φ_P, θ_q and Θ_Q are polynomials in *B* of order *p*, *P*, *q*, and *Q*, respectively, which fulfill the stationarity and invertibility condition.

P and *Q* are the degrees of the autoregressive and moving average seasonal polynomials, Φ and Θ, *p* and *q* are the degrees of the autoregressive and moving average polynomials, respectively. The time-series model requires identification of the functional form of the model, and then the model parameters can be calibrated with sample data sets.

2.2. Identification and validation

The selected model is expected to forecast the next value of ET_0, based on past measurements. The forecast model was obtained by the iterative strategy of specification, estimation, and checking. Thus, the development of a time-series-forecasting model must be done in an iterative fashion, in which (1) the form of the forecasting model is predicted, (2) the coefficients of the model are estimated, and (3) the errors of the forecasting model are analyzed. Steps 1, 2, and 3 are repeated until the errors of the forecasting are reduced to white noise with no significant correlation. More precisely, the time-series model requires the identification of the form of the model, and then the model parameters can be calibrated to the identification sample data.

The values of *p*, *P*, *q*, and *Q* were thus assessed by studying the autocorrelation function (acf) and the partial autocorrelation function (pacf) of the differenced series, and by choosing a SARIMA model where acf and partial acf had a similar form Refs. [24–28]. By analyzing the acf and the partial acf of ET_0, the general form of the forecasting model was developed, then the coefficients of the model estimated and the errors forecasting model were analyzed. The values of *d*, *D*, *p*, *q*, *P*, and *Q* were computed from the Statgraphics-plus software package, whereas the Matlab package was used to develop the computer program for the validation process. From the model structure (0,1,0)(0,1,1)24, developed in Ref. [18], the final value of the prediction was

$$\hat{z}_{k+1} = z_k + z_{k-23} - z_{k-24} + b'_1 \hat{a}_{k-23} \tag{5}$$

where $\hat{z}_{k+1}$ the one-step prediction of the term z_{k+1}.

Eq. (5) expresses the time series as a linear combination of the previous values and the error term. It shows the relationship between the current ET_0 value, past measurements, and error. This model, which describes the evolution of the hourly reference evapotranspiration in the nursery, must be multiplied by the crop coefficient in order to predict the water needs associated with each plot in the nursery.

For the validation purpose, a comparison between the actual and the forecast ET_0 can be seen in **Figure 1** with data sets of two climatic zones. **Figure 1(a)** corresponds to data of Angers 2005, and **Figure 1(b)** to data of Avignon 1999. The climate in the Angers area is typically oceanic (cool and relatively humid summers), with continental influences (wide temperature ranges). The hardest period for the crops (corresponding to maximum water loss) extended from July 9th to July 23rd, 2005. The climate in Avignon is Mediterranean, with dry and hot summer temperatures. The period of greatest evapotranspiration demand for the crops extended from August 15th to August 28th, 1999.

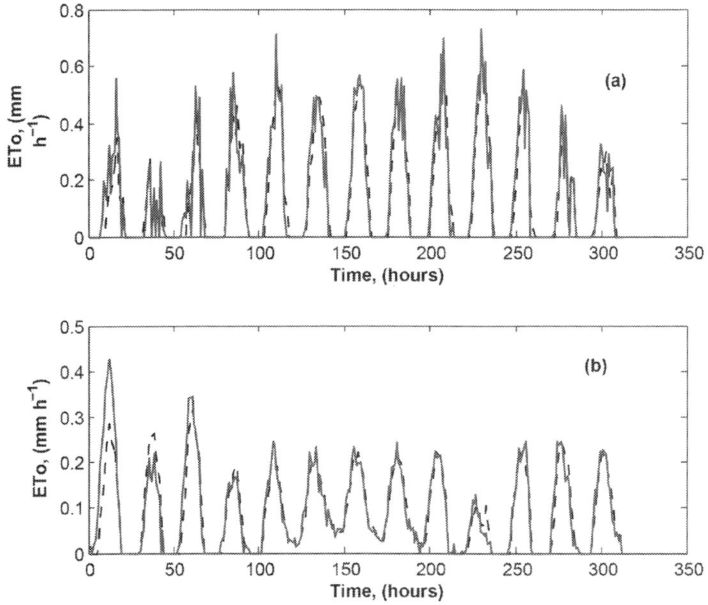

Figure 1. Temporal distribution of hourly ET_0 during validation. ____: Estimated values using PM equation, ……: predicted values from the SARIMA model. (a) Angers July 9th to 23rd, 2005. (b) Avignon, August 15th to 28th, 1999.

One can observe that forecasting errors are most likely within an acceptable range with the average error less than 0.03 mm h$-^1$. The forecasting can be considered accurate despite disturbed weather conditions mainly due to quick variations in net radiation. As seen in **Figure 1**, the hourly predictive model provides a good forecast of the reference evapotranspiration for the two climatic conditions.

3. Predictive irrigation scheduling

3.1. Preliminaries

We briefly recall the nomenclature related to literature on scheduling in computers and manufacturing systems. Machine scheduling considers in general the assignment of a set resources (machines) $M = \{M_1, \ldots, M_m\}$ to a set of jobs $J = \{J_1,\ldots, J_n\}$, each of which consists of a set of operations $J_j = \{O_{j1},\ldots,O_{joj}\}$. The operations O_{jk} typically may be processed on a single machine M_i involving a nonnegative processing time t_{jk}. Usually, precedence constraints are defined among the operations of a job, reflecting its technical nature of processing. Other important aspects that frequently have to be taken into consideration are release dates and due dates of jobs. A solution to the problem is called schedule, assigns start and end times for the operations with respect to the defined constraints of the problem.

Various optimality criteria are based on the completion times C_j of the job J_j in the schedule. The most prominent to mention is the minimization of the maximum completion time (makespan). Another objective can be the minimization of the sum of the completion times. Both measures implicitly attempt to optimize the production costs by minimizing jobs production time. In many situations, due dates d_j which define a required or preferable time of job completion are available for each job J_j. It is then possible to estimate the violations due date in terms of tardiness values T_j. Usual optimality criteria based on this consideration are the minimization of the total tardiness, the minimization of the maximum tardiness, the minimization of the total tardiness or the minimization of the number of tardy jobs.

Scheduling theory covers different models usually specified according to three-field classifi cation $\alpha/\beta/\gamma$. α specifies the machine environment (single-stage systems or multistage systems (covering flow shop, job shop, or open shop problems), β specifies the job characteristics (processing time of job *j* on machine *I*, or released time, due date or weight, etc.), and γ determines the optimality criterion (makespan, total completion time, total weighted completion time, etc.). These problems appear usually to be NP-hard, and are investigated by approximation algorithms, or heuristics algorithms. The scheduling issue when operations durations are known consists of determining depending of the criterion earlier or latest starting time, earlier latest completion time, and so on.

3.2. Scheduling algorithm

Irrigation scheduling has conventionally aimed to achieve an optimum water supply for productivity, with soil water content being maintained close to field capacity. Among the existing irrigation mode trickle, ebb and flow, or sprinkler, the latter is the widely used by growers, and will be considered in the following. The approach remains valuable for other irrigation modes.

A nursery is composed of a set of N plots in which crops at different stages of growth have different water needs. During early stages of growth, plants water need is relatively low, and increases as the plants canopy extends. Therefore, if precise amounts of water have to be applied, grouping plants within zones of irrigation based on containers size and on stage of growth is important. As plants grow, containers are spaced to allow more sunlight penetration and improve plant quality. Containers spacing and canopy characteristics could affect the amounts of overhead water that fall unintercepted between containers, and should be considered in application efficiency evaluation. Difficulties in water management arise when water availability and equipment are insufficient to permanently meet the full crop water requirement. For example, in operating conditions the respect of both allowable pressure head variation of the hydraulic network and of the sprinkler discharge variation in order to ensure emission uniformity leads to the limitation of the number of plots simultaneous irrigable. The main line value needs also to be considered since the sum of distributed discharges should be less than the nominal value. Moreover, the management complexity increases during sunny periods with high values of the probability of simultaneous irrigation requirements by the different species. This can sometimes cause considerable irrigation delays. The aim is to develop a satisfactory water distribution plan and to avoid irreversible damage to production. A priority value can be assigned to each plot in order to preserve the most sensitive crops from water stress in comparison with more resistant crops.

When considering a multiple crops nursery with high-frequency irrigation demand under limited resources, irrigation-triggering scheduling consists of which plot to irrigate, when to irrigate, and the irrigation duration, taking operating constraints, cumulative constraints, and temporal constraints into account. To suitably fulfill these objectives, one should minimize the irrigation starting time, the head pressure losses, and the water stress periods. Referring to the preliminaries above, the problem under consideration differs from the standard parallel-machines-sequencing problem because of its multiobjective aspects and because of the fact that operations and jobs are merged. Thus, instead of using the three-field classification $\alpha/\beta/\gamma$, the problem is formulated in the following compact form:

$$[\, r^{S*}_{c,k+1}, d^{S*}_{c,k+1}] = \arg\min_{c \in U_{ad}} H(SL^{S}_{c,k}, WD^{S}_{c,k+1}, Ae^{S}_{c,k}, p^{S}_{c,k}, Q_{k}, D_{k}) \tag{6}$$

the superscript s stands for growth stage, subscript c for plot or crop, and k for the discrete time.

$r^{s*}_{c,k}$ is the irrigation starting time, $d^{s*}_{c,k}$ the irrigation duration, $SL^{s}_{c,k}$ the water stress level of a crop, $p^{s}_{c,k}$ the priority of a crop, $Ae^{s}_{c,k}$ the application efficiency, Q_k the set of parameters relative to hydraulic constraints, D_k to the scheduling horizon, and $WD^{s}_{c,k}$ to the substrate water deficit.

A simplified hourly soil water balance equation was used to calculate the water deficit. The balance equation was defined as

$$WD^{S}_{c,k+1} = WD^{S}_{c,k} + ET^{S}_{c,*} - I^{S}_{c,k} + R_{k} \tag{7}$$

where $WD^{s}_{c,k}$ (mm) is the soil water deficit at time step k, R_k (mm) is the effective rainfall, $I^{S}_{c,k}$ (mm) is the effective irrigation, and $ET^{S}_{c,*}$ is the crop-specific evapotranspiration: ($\widehat{ET}^{s}_{c,k+1}$ for the predictive scheduling and $ET^{s}_{c,k}$ for the nonpredictive scheduling). The pseudo code of the proposed heuristic based on dispatching rules [29] that prioritize irrigation requests that are waiting for processing is given below:

begin

  for k=0 to 23 **do**

   Compute or forecast ETo

   for c=1 to N **do**

     Estimate the water deficit

     while k'<Kmax **do**

       Compute Uad the set of admissible plots

```
                for l=1 to N do
                    Build the irrigation sequence
                    Update the slack scheduling horizon
                    Update plots priority and water deficit
                endfor
            endwhile
        endfor
    endfor
end.
```

An irrigation request is emitted when the predicted value or the estimated value of the water deficit exceeds the predefined water stress threshold. As the optimization criteria are often conflicting, not a single but a set of solutions are regarded. The resolution of the problem lies in a hierarchical analysis in which subsets of local solutions are derived from the progressive consideration of the constraints. Roughly speaking, the heuristic proceeds by sequential evaluation of the subsets of solutions. The heuristic performs a soil water balance on hourly basis. It calculates or predicts crop ET and then estimates or predicts actual soil water depletion within the root zone. From the priority rules, predicted irrigation dates and amounts are determined based on the current soil water status and anticipated future depletions.

3.3. Simulation results

In order to illustrate the usefulness of the approach based on the predictive scheduling and of the nonpredictive scheduling, a small nursery with 16 plots numbered A1, A2, A3, A4…., D1, D2, D3, D4, was considered, with one crop variety per plot. Plot priority was an integer chosen between 1 and 4, depending on sensitivity of the crop to water stress. The water deficit thresholds were set between 1 and 3.5 mm, while the crop coefficients were chosen ranging from 0.35 to 1. The irrigation water reached the plots through a hierarchical network of main canal, secondary and tertiary canals. The pipeline diameters were fitted in decreasing order of 120 mm for the main, 60 mm for the submain and 45 mm for the sub-submain from the hydrant to the entrance of the plot. The water flow of the main line was fixed at 9 l/s. In order to avert water hammer, water velocity ranged from 0.5 to 2 m/s. From the pressure drop abacus, the maximum water flow per plot was fixed at 2.08 l/s. Considering these hydraulic constraints, only four plots could be simultaneously watered. Application efficiencies ranged from 35 to 80%, depending on canopy development and containers layout on the plots. The software used for the control of irrigation scheduling was written using the Matlab package.

Figure 2 represents results of irrigation scheduling on day 10, with data of the period, July 9th to 30th, 2013, in Angers. The Gantt chart shows only irrigation events between 10 and 19 h. For the nonpredictive scheduling irrigation events arising after 20 h are omitted in order to relieve the representation. For the same reason, irrigation request times or release times are not indicated.

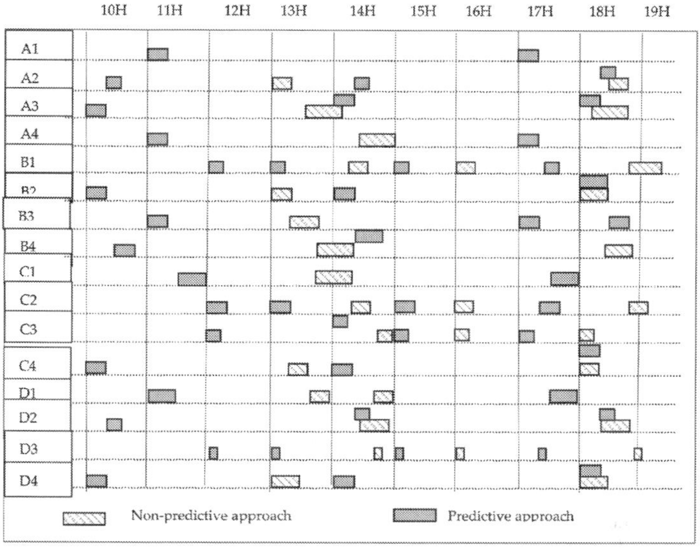

Figure 2. Irrigation scheduling (day 10, Angers, 2013).

One can observe that the irrigation events in the predictive-scheduling approach are more mostly staggered over the working window than those of the nonpredictive scheduling. Indeed, as explained in heuristic description, irrigation requests in the predictive approach are emitted earlier compared to the nonpredictive case. As a consequence, when constraints are satisfied crops are watered before the water-deficit threshold is reached. In this case, crops remain most of the time in a hydric comfort zone. Moreover, doses are reduced leading to a better water sharing or distribution. Irrigation requests are also usually satisfied because of the lower value of the hydraulic network load on the scheduling horizon, and the use of the rules prioritizing the requests. For this simulation, there is no significant bottleneck. A peak water requirements period appears at hour 17. Irrigation events are spread over the scheduling horizon without calling into question the capacity of the hydraulic network. The approach can be considered as sustainable since the amount of water required by the crop is applied at the proper timing to prevent the soil water content from becoming dryer than the management allowable depletion.

In the nonpredictive case, two bottlenecks can be observed as a consequence of sudden high evaporative demands. For the first period 12–14 h, irrigation requests are emitted simultaneously and the algorithm produces the given schedule. There is no idle time. One could deduce that the hydraulic network is well designed, since the irrigation requests dates are not represented. As the demand is greater than the main line, priority is given to the plots satisfying the imposed constraints. For example, between hours 13 and 14, request of plot A1 with weak priority value is postponed to the next scheduling horizon. A similar behavior is observed with plots A1 and C2 between hours 19 and 20. Between hours 12 and 13, the irrigation on the plots B4 and C1 spans on two consecutive periods, without preemption due to their highest priority value compared to that of A4, B1, and D1. The second bottleneck period 18–20h presents some similarities with the first period. Many irrigation events are delayed compared to the cases of predictive scheduling. The major drawback of this approach is that the decision to irrigate is made after the plant has suffered some amount of water stress.

In general, one can observe that the recovery of crop water status is rapid and water status is significantly better in the predictive scheduling than in the nonpredictive scheduling.

4. Conclusion

In this study, we proposed a predictive approach of irrigation scheduling in nurseries, with the objectives of minimizing the water stress periods of the crops while optimizing the use of disposal materials. The time-series theory enabled to obtain good forecast of the potential evapotranspiration allowing the prediction of crops evapotranspiration. A heuristic of irrigation sequencing was developed and was applied on a small nursery designed for the purpose. Simulation results with predictive and nonpredictive scheduling showed the ability of the predictive-scheduling approach to proper timing the amount of water required by crops in order to prevent water stress periods which may adversely affect the crop yield. Extension of the approach to sudden changes in weather conditions is under study in order to improve the prediction capability of the heuristic.

Author details

Rousseau Tawegoum[1]*, Florian Leroy[1], Gérard Sintes[2] and Gérard Chassériaux[1]

*Address all correspondence to: rousseau.tawegoum@agrocampus-ouest.fr

1 Agrocampus Ouest – Centre of Angers, Environmental Physics and Horticulture (EPHor) Research Unit, Angers, France

2 Reseach Institute on Horticulture and Seeds, Beaucouze, France

References

[1] European Parliament. Directive 2000/60/EC of the European parliament and the council of 23 October 2000, establishing a framework for community action in the field of water policy. Official Journal. 2000;**L327**. 93p.

[2] Wanjura DF, Upchurch DR. Time threshold for canopy temperature-based irrigation. In: Camps CR, Sadler EJ, Yoder RE, editors. International Conference on Evapotranspiration and Irrigation Scheduling; Nov 3–6, 1996; San Antonio (Texas). 1996. pp. 295–303.

[3] Miranda FR, Yoder RE, Wilkerson JB, Odhiambo LO. An autonomous controller for site-specific management of fixed irrigation systems. Computers and Electronics in Agriculture. 2005;**48**:183–197.

[4] Caceres R, Casadesus J, Marfa O. Adaptation of an automatic irrigation control tray system for outdoor nurseries. Biosystem Engineering. 2006;**96**(3):419–425.

[5] Vellidis G, Tucker M, Perry C, Kvien C, Bednarz C. A real-time wireless smart sensor array for scheduling irrigation. Computers and Electronics in Agriculture. 2008;**61**:44–50.

[6] Cardenas-Lailhacar B, Dukes MD. Precision of soilmoisture sensor irrigation controllers under field conditions. Agricultural Water Management. 2010;**97**:666–672.

[7] Moriana A, Fereres E. Plant indicator for scheduling irrigation of young olives trees. Irrigation Sciences. 2002;**21**:83–90.

[8] Huguet JC, Li SH, Torendeau JY, Pelloux G. Specific micrometric reactions of fruits threes to water stress and irrigation scheduling automation. Journal of Horticultural Sciences. 1992;**67**:631–640.

[9] Goldhamer DA, Fereres E, Mata M, Girona J, Cohen M. Sensitivity of continuous and discrete plant soil water status monitoring in peach trees subjected to deficit irrigation. Journal of American Society Horticultural Sciences. 1999;**124**:437–444.

[10] Jensen MC, Middleton JE, Pruitt WO. Scheduling irrigation from pan evaporation. Washington Agricultural Experimental Station. 1961;**Circular No. 386**.

[11] Saruwati N, Yomota A. Forecasting system irrigation water on paddy field by fuzzy theory. Agricultural Water Management. 1995;**14**:163–178.

[12] Theiveynathan S, Benyoun RG, Marcar NE, Myers BJ, Polglase PJ, Falkiner RA. An irrigation schedule model for application of saline water to tree plantation. Forest Ecology and Management. 2004;**193**:97–112.

[13] Phene RC. Real-time irrigation scheduling with automated evapotranspiration pan system. In: Camps CR, Sadler EJ, Yoder RE, editors. International Conference on Evapotranspiration and Irrigation Scheduling; Nov 3–6, 1996; San Antonio (Texas). 1996. pp. 1093–1098.

[14] Davis SL, Dukes MD. Irrigation scheduling performance by evapotranspiration-based controllers. Agricultural Water Management. 2010;**98**:19–28.

[15] Chartzoulakis. K. Sustainable water management in agriculture under climate change. In: JRC, editor. JRC Conference Scientific Support to Agriculture: Competitiveness, Quality and Sustainability; April 23; Athens (Greece). JRC; 2014. p. 16.

[16] Sigrimis N, Arvanitis KG, Pasgianos GD, Ferentinos K. Hydroponics water management using adaptive scheduling with an on-line optimizer. Computers and Electronics in Agriculture. 2001;**31**:31–46.

[17] Yeager T, Million J, Larsen C, Stamps B. Florida nursery best management practices: past, present and future. Hort Technology. 2010;**20**(1):82–88.

[18] Tawegoum R, Leroy F, Sintes G, Chassériaux G. Forecasting hourly evapotranspiration for triggering irrigation in nurseries. Biosystems Engineering. 2015;**129**:237–247.

[19] Allen RG, Pereira LS, Raes D, Smith M. Crop evapotranspiration: Guide-lines for computing crop requirements. Irrigation and drainage; paper No. 56 FAO. Rome (Italy): 1998. 300 p.

[20] Marino MA, Tracy JC, Taghavi AS. Forecasting of reference crop evapotranspiration. Agricultural Water Management. 1993;**24**(3):163–187.

[21] Hess T. A microcontroller scheduling program for supplementary irrigation. Computers and Electronics in Agriculture. 1996;**15**:233–243.

[22] Duce P, Snyder RL, Spano D. Forecasting reference evapotranspiration. In: Ferreira MI, Jines HG, editors. Third International Symposium on Irrigation Horticultural Crops; June 28–July 2; Estoril (Lisbon) Portugal. 1999. pp. 135–141.

[23] Seity Y, Brousseau P, Malardel S, Hello G, Bernard P, Bouttier F, Lac C, Masson V. The AROME-France convective-scale operational model. Monthly Weather Review. 2011;**139**:976–991.

[24] Box GEP, Jenkins GM, Reinsel GC. Time Series Analysis, Forecasting and Control. 3rd ed. United States of America: Englewood Cliffs, Prentice-Hall; 1994. 598 p.

[25] Akaïke H. A new look at the statistical model identification. IEEE Transactions on Automatic Control. 1994;**19**(6):716–723.

[26] Stoïca P. On a procedure for testing the order of a time series. IEEE transactions on Automatic Control. 1981;**26**(2):562–573.

[27] Gersch W. Estimation of the autoregressive parameters of amixed autoregressive moving average time series. IEEE Transactions on Automatic Control. 1980;**15**(6):585–588.

[28] Purat B. Some asymptotic properties of the sample covariances of Gaussian autoregressive moving average process. Journal of Time Series Analysis. 1987;**8**(2):205–220.

[29] Leroy F. Simulation of irrigation piloting in nurseries: estimation of water requirement and control of hydraulic network.[Master dissertation on Automatic Control], University of Angers (France):1999. 45 p.

Chapter 3

Municipal Wastewater Irrigation for Rice Cultivation

Dong Duy Pham and Toru Watanabe

Additional information is available at the end of the chapter

http://dx.doi.org/10.5772/66956

Abstract

In scene of worrisome water shortage, municipal wastewater has been gradually accepted as an alternative water resource containing important nutrients for irrigation. Rice cultivation, which is one of the main crops feeding global population and requires plenty of water for its effective growth, has been often irrigated by municipal wastewater in many countries. While irrigation of municipal wastewater for rice cultivation must bring benefits for farmers mainly by increased yield with less amount of fertilizers, it also has potential to cause drawbacks to human health and the environment. This chapter discusses about these aspects based on scientific works and practical experiences of municipal wastewater irrigation for rice production as well as the introduction of our concept to cultivate rice for animal feeding with irrigation of treated wastewater, which can contribute to resource circulation between urban and rural areas. The feasibility study under this concept has demonstrated that the target value of rice yield can be achieved and protein-rich rice preferable for animal feed can be harvested with irrigation of properly treated municipal wastewater.

Keywords: municipal wastewater, rice production, wastewater irrigation, benefit and disadvantage, greenhouse gas, rice for animal feeding

1. Introduction

Climate change and global population explosion put water resources scarcity in many corners of the world to alarming status [1, 2], with around 1.1 billion people lacking access to freshwater in developing countries and nearly 2.4 billion lacking adequate sanitation [3]. It is estimated that two-thirds of the world's population will suffer from moderate to high water stress, and about half of the population will face severe water supply constraints in 2025 [4]. Agriculture is known as the largest consumer of freshwater resources at the beginning of the twenty-first century, and water consumption for agriculture accounts for over 70% of global water withdrawals [5, 6]. However, agricultural irrigation water does not usually require the same high grade of water quality as drinking water [7]. Currently, in approximately 1.5 billion hectares of agricultural land all over the world [8], total fertilizer use ($N + P_2O_5 + K_2O$) is estimated to be around 190.4 million tons [9]. Wastewater can supply a significant amount of nutrients which can improve soil fertility, plant growth and crop production, reducing the consumption of required fertilizers [10]. In this circumstance, municipal wastewater is evaluated as a new resource of water, and the practice of reclaimed wastewater for agricultural irrigation is likely to become more commonly applied in many countries with a vast volume [11–13]. It is not the main objective, but reuse of wastewater also contributes to interrupting discharges of nutrients and organic matters into water environment [14].

Approximately 50% of worldwide irrigation water is used by rice cultivation—one of the agricultural products which need notable amount of water [15, 16]. Although there are a variety of practices of rice cultivation, typically, rice fields are flooded before plowing, and thereafter, water levels are kept at 4–6 cm in shallow rice fields. It sometimes becomes as high as 10 cm when continuous-flooding irrigation is done during the growing season [17]. Rice is one of the leading cereal crops providing around 20% daily calories for more than 3.5 billion people [18]. There are around 150 million hectares of rice land worldwide, which supply 550–600 million tons of rough rice yearly. Irrigated rice cultivation makes up 55% of harvested rice area and contributes to 75% of global rice production [19]. Irrigation of rice paddy using treated or untreated wastewater is extensively practiced and examined in many countries to investigate benefits [20–26] or drawbacks of the practices [12, 27–33].

This chapter provides information about municipal wastewater characteristic as well as discussions about positive and negative effects of its reuse for rice cultivation.

2. Municipal wastewater and its treatment

2.1. Characteristics of municipal wastewater

Municipal wastewater, which is usually conveyed in a combined sewer or sanitary sewer, consists of domestic wastewater, industrial wastewater, and storm water and groundwater seepage entering the municipal sewage network. Domestic wastewater includes effluent from households, institutions, commercial buildings and the like. Industrial wastewater is the effluent discharged from manufacturing units and food processing plants. In general, characteristics of domestic wastewater are not significantly different from one region to another, while there are many types of industrial wastewater based on industrial processes as its origin.

Municipal wastewater mainly consists of water (99.9%) together with relatively low concentrations of suspended and dissolved organic and inorganic solids. Parts of the organic substances present in wastewater are carbohydrates, lignin, fats, soaps, synthetic detergents, proteins and their decomposition products as well as various natural and synthetic organic chemicals from the process industries. **Table 1** shows the levels of the major constituents in municipal wastewater.

Contaminants	Unit	Range
Total solid (TS)	mg/L	390–1230
Total dissolved solid (TDS)	mg/L	270–860
Total suspended solid (TSS)	mg/L	120–400
Biochemical oxygen demand (BOD_5)	mg/L	110–350
Chemical oxygen demand (COD)	mg/L	250–800
Total organic carbon (TOC)	mg/L	80–260
Total nitrogen (TN)	mg/L	20–70
Total phosphorus (TP)	mg/L	4–12
Total coliform	no./100 mL	10^6–10^9
Fecal coliform	no./100 mL	10^3–10^7

Sources: [34].

Table 1. Typical composition of untreated domestic wastewater.

2.2. Treatment of municipal wastewater

It is not recommended to reuse municipal wastewater directly for rice cultivation due to its drawbacks, which are described in the next section. Treatment of wastewater at any level is required to overcome the drawbacks. The principal objective of wastewater treatment is to remove contaminants such as solids, organic matter and nutrients before the treated wastewater is discharged into water bodies. The quality of treated wastewater depends on the treatment technology and operation.

Although wastewater treatment includes physical, chemical and biological processes, it normally has four basic steps: preliminary, primary, secondary and advanced treatments [35]. Preliminary treatment is designed to remove coarse solids and other large materials, which are often found in raw wastewater. These solids consist of pieces of wood, cloth, paper, plastics, sand, gravel, etc. The objective of primary treatment is to extract organic and inorganic solids from wastewater by the physical process of sedimentation and flotation. Approximately 25–50% of the BOD, 50–70% of the SS and 65% of the oil are removed throughout this treatment step [36].

Secondary treatment, in general, follows primary treatment to do the further treatment. Its objective is removal of biodegradable dissolved and colloidal organic matters from effluent of primary treatment using many different types of microorganisms in a controlled environment. The principal secondary treatment techniques are the trickling filter and the activated sludge process. The latter one, which is used most commonly all over the world, can remove organic matters effectively but cannot do nutrients, especially nitrogen, from wastewater. Hence, the secondary effluent from wastewater treatment plants still has a high content of nutrients available for crop growth.

At most treatment plants, the secondary effluent is discharged into receiving water environment after disinfection with chlorine, ozone or ultraviolet radiation. To prevent eutrophication in the water environment, advanced treatment is sometimes applied to remove specific contaminations such as nutrients in the secondary effluent [37].

2.3. Advantages and disadvantages of wastewater irrigation for rice production

These characteristics of municipal wastewater make us imagine advantages and drawbacks of its irrigation for rice production. Major advantages are:

- Higher crop yields with reduced use of synthetic fertilizers, resulting in saved cost for cultivation.
- Enhanced recycles of nutrients and organic matters, improving soil properties.
- Reduced discharges of pollutants to surface water bodies.
- Decreased freshwater withdraw during irrigation.

On the other hand, we should pay attentions to its drawbacks such as:

- Contamination of irrigated soil with salt, heavy metals and toxic compounds originated from wastewater, resulting in reduced soil productivity.
- Contamination of agricultural products (rice crop) with heavy metals and toxic compounds, posing health risks to consumers.
- Farmers' risk of health problems due to exposure to paddy water contaminated with pathogens, heavy metals and toxic compounds.
- Contamination of groundwater due to infiltration of wastewater used for irrigation.

The following sections describe more detail explanation about the above advantages and drawbacks. Most of them are common to irrigation of the treated wastewater, although its treatment may highlight the advantages and overcome the drawbacks.

3. Potential impacts of municipal wastewater reuse for rice production

3.1. Impacts on crops

In general, wastewater irrigation can affect rice crops in terms of yields and crop quality such as appearance and flavor. Municipal wastewater is a rich source of nutrients necessary for crop growth, so it is expected that crops irrigated with municipal wastewater get higher yield than normal. Yoon et al. [20] examined the effect of treated sewage irrigation on paddy rice cultivation. They found that the irrigation did not adversely affect the growth and yield of rice, resulting in up to 50% greater yield than rice without wastewater irrigation. Thu [38] also reported that wastewater irrigation brought 10–15% higher yield of rice crops.

If nitrogen supplied to the crop exceeds its dose recommended for optimal yields, crop growth may be stimulated together with yield loss and delayed ripening [14]. The study by Nyomora [26] illustrated that wastewater irrigation resulted in four times higher rice yield than tap water irrigation, but, in contrast, wastewater irrigation applied with N-P-K fertilizer depressed the yield potential to 3.2 times of that obtained without its application. This situation can happen accidentally. For example, urea plant effluents, as a rich source of liquid fertilizer in concentrated forms, have adverse effects on rice and corn yields [2]. Also, oversupply of nitrogen may be resulted in overgrowth of rice plants, which triggers their lodging and reduces eating quality of rice due to increased content of proteins [39].

Crops irrigated with wastewater have potential to be contaminated with microbes, heavy metals and organic toxic compounds in wastewater. This effect is discussed separately in the Section 3.5.

3.2. Effects on soil resources

Wastewater can affect paddy soil in two opposite ways: by providing benefits and causing problems. It is usually difficult to predict which effect appears in wastewater irrigation because soil is a very complicated structure involving inorganic and organic matters. One of the most recognizable effects of wastewater irrigation is a rise of yield due to nutrients supplied with wastewater as well as soil texture improved by organic matters in wastewater [40]. Supplying organic matter improves soil texture by enhancing soil humidity and microbial activity [41].

Nitrogen in wastewater consists of several chemical forms such as nitrate, nitrite, ammonia and organic nitrogen. All of these forms are soluble and mobile in water, and when the wastewater is irrigated, all forms of nitrogen except ammonia are easily washed out and may cause pollution of groundwater and surface water receiving the runoff water. Only ammonia in wastewater can attach to soil particles and is retained in paddy fields, but, at the surface of soil layer and rhizosphere with the presence of oxygen, it is gradually converted to nitrite and finally nitrate with bacterial activities. By contrast, phosphorus, which can exist as a trivalent cation, is so stable in soil layer. In addition to this fact, since wastewater contains a smaller amount of phosphorus than that required by crops, its irrigation hardly gives an adverse impact on the water environment [14].

On the other hand, wastewater irrigation may make consequent adverse effects on soils. The most commonly reported impact is accumulation of metals that, depending on the level, may be harmful. Chung et al. [12] indicated that application of domestic wastewater to arable land for 3 years slightly increased the levels of Pb, Cd, Cu and Zn in soil. Kang et al. [21] conducted rice cultivation with groundwater and treated wastewater in different treatment levels. The results showed no adverse effects on chemical concentrations including the heavy metals (Cu, As, Cd, Zn, Hg and Pb) in paddy soil. This indicates the possibility that treated municipal wastewater can be safely used as an alternative water source for the irrigation of rice, although continued monitoring will be needed to determine the long-term effects with regard to soil contamination.

A field research in Thessaloniki, Greece, during a 2-year period [22] reported no adverse effects on the physicochemical properties of soil, whereas macro and trace elements concentration showed discrepancies between the 2 years and the three treatments (river water with N-P fertilizer, treated wastewater with N fertilization and treated wastewater without fertilizer).

Wastewaters including industrial discharges with a high metal concentration are harmful to crops and eventually to consumers, as a result of metal accumulation in soil. The elements of major concern are heavy metals such as cadmium, copper, molybdenum, nickel and zinc. Yang et al. [27] reported that the paddy soil irrigated with untreated mining wastewater in Lechang lead/zinc mine area was heavily contaminated by Cd and would pose a human/animal health risk through Cd mobility in the food chain. Very high concentrations of As, Cd, Cu, Pb and Zn were found in the paddy soils irrigated by river water, which received wastewater from mining activity [42].

Wastewater, particularly domestic wastewater, normally contains salts which may be accumulated in the root zone with possible harmful impacts on soil health. Increase rate of salinity depends on the salinity of irrigated water, soil transmissivity, organic matter concentration, land drainage, irrigation rate, depth to the groundwater level and the type of soils. Long-term use of wastewater with high salt contents is a potential hazard for soil as it may erode the soil structure, resulting in less productivity. The problem of soil salinity can be settled by the application of natural or artificial solutions, although it is costly and leads to economic constraints.

Wastewater with a large amount of solids may cause soil clogging, depending on soil porosity, concentration (>100 mg/L can cause the problem) and chemical composition. The most concerning components are minerals that are not biodegraded. If soil is clogged, irrigation will become less effective due to dismissed water percolation [43].

3.3. Effects on ground and surface water

The first effect of irrigated agriculture on groundwater resource is aquifer recharge. The recharge happens almost always non-intentionally and has the advantage of increasing the local availability of water [44]. Pescod [36] estimated that 50–70% of the irrigation water could infiltrate to groundwater aquifer in some parts.

Due to this phenomenon, wastewater irrigation can cause adverse effects on groundwater resource. The most famous adverse effect is infiltration of nitrates in irrigated wastewater into groundwater. Groundwater contaminated with nitrates is known to cause methemoglobinemia in infants, so-called blue baby syndrome, if it is used as a source of drinking water [43].

Not only nitrogen but also organic matters and metals may contaminate groundwater in municipal wastewater irrigation. If some of most toxic metals to humans—cadmium, lead, mercury and arsenic—are present in irrigated wastewater at a higher concentration than the acceptable level, groundwater is severely contaminated, posing risk of serious diseases like cancer to the groundwater users. Contamination of groundwater with organic matters brings another type of health risk to its users, through the formation of organochlorides when the groundwater is disinfected with chlorine (the most common method) for drinking purpose [45].

Long-term irrigation of municipal wastewater may result in a significant increase of salt content in aquifers, although quality of irrigated wastewater, soil characteristics and original quality of the receiving groundwater are all important factors to determine the extent to which the quality of groundwater is impacted. Even though groundwater has a low salt concentration, addition of salts originated from irrigated wastewater may not be considered too adverse if its movement is limited or if it is not used for any purposes. Thus, the impact of increased salts in groundwater by wastewater irrigation, which is sometimes inevitable, needs to be weighed up in consideration with all the risks and benefits from the irrigation [46].

Surface water bodies are also affected due to drainage and runoff from the fields irrigated with municipal wastewater. The inevitable contamination in surface water is almost the same as that in groundwater, but the extent of the impact depends on the strength of wastewater and type of water body (i.e., river, irrigation channel, lake or dam) as well as hydraulic retention time in the fields.

3.4. Effects on quality of irrigated wastewater

Although wastewater irrigation has a potential to contaminate freshwater sources, it is expected that quality of the wastewater is improved by being used for irrigation. Suspended solids including pathogenic microorganisms are trapped and absorbed in upper soil layers and removed from the wastewater. The efficiency of solid removal depends on sizes of soil pore and the solids [44]. Adsorption of microorganisms to soil particles is favored at low pH, high salt concentration in the sewage and high relative concentrations of calcium and magnesium over monovalent cations such as sodium and potassium in soil [14].

Organic matters in wastewater can be rapidly converted in soils to stable and nontoxic ones such as humic and fulvic acids. In fact, we can find biodegradation of a wider variety and greater amount of organic matters in soils than in water bodies. So the organic matters in term of COD and BOD in the irrigated wastewater are significantly decreased after percolation through soil layers.

More significant reduction in nitrogen concentration is expected at paddy fields with wastewater irrigation due to three main reasons: absorption by plants, release to the atmosphere as the result of nitrification and denitrification by nitrogen bacteria such as *Nitrobacter* and *Nitrosomonas*, and adsorption of ammonium to soil particles. Firstly, rice plants grow taking nutrients in wastewater used for irrigation, and nitrogen, one of the fundamental nutrients for plant development, is removed from the wastewater stored in soil layers [15, 24]. Secondly, soil and rice rhizosphere microorganisms contribute to transformation of organic nitrogen or ammonium to nitrogen gas as well as nitrous oxide gas under a variety of redox conditions in soil layers [23]. Nitrogen removal is enhanced if flooding and drying periods are alternated for promoting nitrification/denitrification process, with 75% removal at the maximum [14]. Thirdly, ammonium as a cation has an affinity to the surface of soil particles normally with positive charge. However, a large amount of ammonium is supplied, and as mentioned above, excess nitrogen will be transported to groundwater with infiltrated irrigation water. Nitrites and nitrates, which are anions, are easily lost from paddy fields, resulting in groundwater contamination.

3.5. Effects on human health

As mentioned above, municipal wastewater includes pathogenic microorganisms such as bacteria, viruses and parasites. These microorganisms potentially pose human health risks when the wastewater is reused for some activities. Particularly, human parasites such as protozoa and helminth eggs are of special significance in this concern as they are known as being more difficult to remove by treatment processes [2].

Paddy fields irrigated with municipal wastewater may have unfavorable health effects on farmers. It has been reported that the practice of reuse of raw or even treated wastewater for irrigation may cause epidemiological problems among nearby populations and consumers of uncooked agricultural products [47]. The degree of risk may vary among the various age groups [2], and in a study [31], children were found to have a greater risk of infection with *Escherichia coli*. A study conducted in a province in northern Vietnam [29] assessed the risk of skin disease among farmers occupationally exposed to wastewater, showing that exposure to wastewater is a major risk factor for skin disease, but it is not clear which chemical and biological agents might play the main role in causing the diseases. Rhee et al. [30] examined the concentrations of *E. coli* in a paddy rice field irrigated with reclaimed wastewater and evaluated the risk of its infection among farmers using beta-Poisson dose-response model. The results showed that the risk was lower in irrigation of groundwater and reclaimed wastewater irrigation than in irrigation of direct effluent from wastewater treatment plant.

Municipal wastewater sometimes has harmful metals such as Zn, Cu, Pb, Mn, Ni, Cr and Cd, depending upon the type of activities in the associated area. Continuous irrigation of municipal wastewater may result in heavy metal accumulation in the soil and agricultural products [48]. In case of rice plant, it is well known that Cd is the metal to which a special attention should be paid because it is accumulated so intensively in edible part of rice.

Most of the heavy metals are normally removed well by wastewater treatment processes. Even so, we should take case about heavy metal contamination in the paddy field considering subsequent food chain involving agricultural products and consumers [49]. Due to the nonbiodegradable and persistent nature, heavy metals are accumulated in viscera and born, and are associated with numerous serious health disorders [50]. Singh et al. [51] indicated that rice and wheat grains contained less heavy metals than vegetables, but health risk was more significant due to higher contribution of cereals in the diet.

3.6. Socioeconomic effect

Wastewater irrigation brings various economic benefits. Wastewater for irrigation does not require as high quality as the effluent which is discharged to water bodies. Indeed, thanks to the function of paddy fields to improve water quality as explained in the Section 3.4, the discharge from the field has a better quality than the irrigation water. By using this function effectively, we can save the cost for wastewater treatment.

In addition, when wastewater containing rich nutrients is used for irrigation, we can reduce the amount of fertilizer applied to the field, resulting in cost saving or higher yield obtained.

http://dx.doi.org/10.5772/66956

This must contribute to improvement of economic status of farmers. Papadopoulos et al. [22] reported that the total production cost decreased up to 11.9% by applying municipal wastewater for rice production, compared to the normal paddy field.

3.7. Effects on greenhouse gas emission

Global warming is caused by the emission of greenhouse gases (GHGs) such as methane (CH_4) and nitrous oxide (N_2O). On global scale, agricultural activities accounted for about 50% of CH_4 and 60% of N_2O in the total anthropogenic GHGs emissions in 2005 and nearly 17% increase of these emissions from 1990 to 2005 [52]. In particular, paddy fields and irrigated lowland rice production systems are known to be significant sources of CH_4 and N_2O, which are two important trace gases contributing to an observed increase of approximately 0.6–0.7°C in global surface temperature during the last century.

GHGs emission from paddy fields may be affected by many factors such as water regime, organic matter and nitrogen resource including fertilizer. As introduced above, municipal wastewater is rich in organic matters and also contains an appreciable amount of macronutrients and micronutrients, and thus nutrient levels of soils are expected to increase with its irrigation. Several studies focused on the effects of water regime and fertilizer application on GHGs emission strength; however, to our knowledge, there was only one research [53] examining the effect of wastewater irrigation on CH_4 and N_2O emissions from paddy field. Reports showed that CH_4 and N_2O emissions from rice paddies are closely associated with soil carbon and nitrogen availabilities and transformation processes, which are significantly dependent on soil properties, soil heavy metal contents and soil microbial communities [54–56]. Consequently, Zou et al. [53] hypothesized that wastewater irrigation would significantly increase these gas emissions from rice paddies. The increments of CH_4 and N_2O emissions were 27 and 68%, respectively, compared to paddy fields irrigated by river water.

4. New concept: cultivation of rice for animal feeding with irrigation of municipal wastewater

4.1. Motivation and goal

In previous sections, we discussed the benefits and downsides of municipal wastewater reuse for irrigation, in particular for rice cultivation. Most of the drawbacks are from the contaminants in the irrigated wastewater, and therefore, one of the best ways to reduce its adverse effects is to use treated wastewater after the contaminants are removed to the suitable level. For this purpose, advanced treatments are not necessary and low-cost technologies are preferable to keep the total cost for cultivation acceptable.

Such low-cost technologies, even standard activated sludge process, are difficult to remove nutrients from wastewater, and the application of treated wastewater may lead to overgrowth of rice plants, resulting in lodging [15]. Too much supply of nutrients, especially nitrogen, also reduces the eating quality of rice due to high content of proteins.

These difficulties in irrigation of treated wastewater gave us motivation to propose a new concept to cultivate rice for animal feed rather than for human consumption [39]. The rice cultivars used for animal feed have several advantages compared with those used for human consumption. These advantages include higher crop yield and plant resistance to lodging. Moreover, the high protein content in this rice, which results from the adsorption of excess nitrogen, is preferable for animal feed.

We can expect that rice cultivation in this concept contributes to an improvement in the quality of treated municipal wastewater. In addition to this merit in environmental protection, it is also expected to promote water and nitrogen circulation among urban dwellers who consume animal products and produce wastewater, farmers who produce rice for animal food by reusing treated wastewater and livestock farmers who use the cultivated rice as fodder for the animals (**Figure 1**). Our goal is to realize this resource circulation based on the new concept for sustainability of our environment. We recognize that it is another advantage of this concept to overcome the psychological hurdle of consumers against eating rice cultivated using treated wastewater.

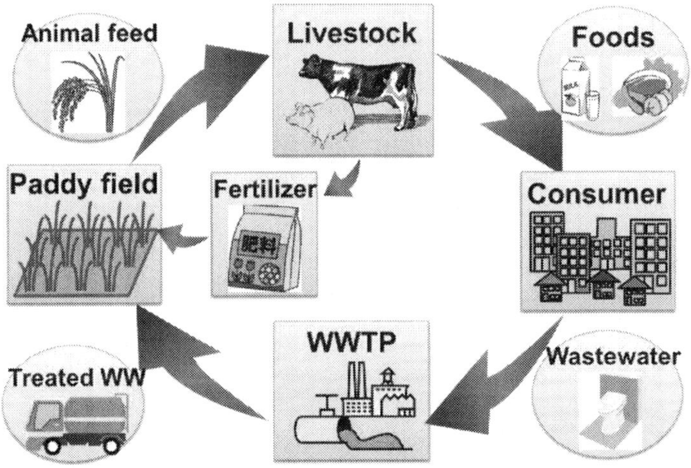

Figure 1. Resource circulation involving urban (consumers) and rural areas (rice and livestock famers), which is realized with cultivation of rice for animal feeding with irrigation of treated municipal wastewater.

4.2. Progress in our research toward implementation of this concept

The bench-scale experiment (**Figure 2**) with dimension of 0.6 × 0.3 m revealed that cultivation of rice for animal feeding could remove three times the amount of nitrogen from the treated wastewater compared with rice cultivation for human consumption [39]. In addition, the experiment showed that upward irrigation called bottom-to-top irrigation, in which treated wastewater is supplied from the pipe fixed at the bottom of the field and then infiltrate up through soil layer to the surface, increased the nitrogen released to the atmosphere, probably

http://dx.doi.org/10.5772/66956

Figure 2. Bench-scale experiment to reveal the performance of treated wastewater irrigation to cultivate rice for animal feeding.

because of enhanced denitrification. This kind of irrigation seemed to increase the rice yield and biomass of the plants.

The yield of rice reached its target (8t/ha) for the cultivar used for the experiment [57], and the protein content (up to 13.1%) in the rice cultivated with irrigation of treated wastewater was significantly higher than that found in the normal paddy fields [58]. Actually, it may be possible to harvest such a protein-rich rice if a larger amount of nitrogen fertilizer is applied, but it is not cost-effective and attractive to farmers. In this sense, application of treated wastewater, which enables low-cost cultivation, is essential and core in the concept. We are now trying to examine the performance of rice cultivation practice, which was revealed in the bench-scale experiment, in the real fields.

5. Conclusions

Rice, which is a leading cereal crop, demands a huge amount of water, while the exploding urban population needs foods as well as produces wastewater. Reuse of wastewater for rice cultivation has a great potential to contribute to sustainable wastewater management. Several instances of positive and adverse effects of municipal wastewater irrigation for rice cultivation were given in this chapter. To avoid such negative effects on environment and human health, it is highly recommended that municipal wastewater should be reused for irrigation after being treated properly. Supposing it is treated with activated sludge process, a new concept "rice cultivation for animal feeding with irrigation of treated municipal wastewater" was introduced. Until now, our bench-scale experiment has revealed the feasibility of this concept with the data showing the achievement of target value of rice yield and its high protein content which is preferable for animal feed.

Acknowledgement

This review was based on the outcome of our researches supported by the Ministry of Land, Infrastructure, Transport, and Tourism (MLIT) of Japan through the Gesuido Academic Incubation to Advanced (GAIA) project, by the Ministry of Education, Sports, Culture, Science, and Technology (MEXT) of Japan through the Center of Community (COC) project and by the Institute for Regional Innovation, Yamagata University.

Author details

Dong Duy Pham[1] and Toru Watanabe[2*]

*Address all correspondence to: to-ru@tds1.tr.yamagata-u.ac.jp

1 The United Graduate School of Agricultural Sciences, Iwate University, Morioka, Japan

2 Faculty of Agriculture, Yamagata University, Yamagata, Japan

References

[1] N. W. Arnell. Climate change and global water resources: SRES emissions and socio-economic scenarios. Global Environmental Change. vol. 14, no. 1, pp. 31–52. 2004.

[2] I. Hussain, L. Raschid, M. A. Hanjra, F. Marikar and W. van der Hoek. Wastewater use in agriculture: Review of impacts and methodological issues in valuing impacts. Working Paper 37. International Water Management Institute, Colombo, Sri Lanka. 2002.

[3] S. P. Simonovic. World water dynamics: global modeling of water resources. Journal of Environmental Management. vol. 66, no. 3, pp. 249–267. 2002.

[4] V. Lazarova, B. Levine, J. Sack, G. Cirelli, P. Jeffrey, H. Muntau, M. Salgot and F. Brissaud. Role of water reuse for enhancing integrated water management in Europe and Mediterranean countries. Water Science and Technology. vol. 43, no. 10, pp. 25–33. 2001.

[5] UNESCO. Agriculture, food and water: A contribution to the World Water Development Report. 2013.

[6] S. H. Gheewala, T. Silalertruksa, P. Nilsalab, R. Mungkung, S. R. Perret and N. Chaiyawannakarn. Water footprint and impact of water consumption for food, feed, fuel crops production in Thailand. Water (Switzerland). vol. 6, no. 6, pp. 1698–1718. 2014.

[7] T. Jang, M. Jung, E. Lee, S. Park, J. Lee and H. Jeong. Assessing environmental impacts of reclaimed wastewater irrigation in paddy fields using bioindicator. Irrigation Science. vol. 31, no. 5, pp. 1225–1236. 2013.

[8] FAO. World agriculture: towards 2015/2030—An FAO perspective. p. 432. 2003.

[9] FAO. Current world fertilizer trends and outlooks to 2015. 2011.

[10] M. A. Hanjra, J. Blackwell, G. Carr, F. Zhang and T. M. Jackson. Wastewater irrigation and environmental health: implications for water governance and public policy. International Journal of Hygiene and Environmental Health. vol. 215, no. 3, pp. 255–269. 2012.

[11] UNEP. Water and Wastewater reuse: An Environmentally Sound Approach for Sustainable Urban Water Management. 2005.

[12] B. Y. Chung, C. H. Song, B. J. Park and J. Y. Cho. Heavy metals in brown rice (*Oryza sativa* L.) and soil after long-term irrigation of wastewater discharged from domestic sewage treatment plants. Pedosphere. vol. 21, no. 5, pp. 621–627. 2011.

[13] D. Norton-Brandão, S. M. Scherrenberg and J. B. van Lier. Reclamation of used urban waters for irrigation purposes: a review of treatment technologies. Journal of Environmental Management. vol. 122, pp. 85–98. 2013.

[14] B. Jiménez. Irrigation in developing countries using wastewater. International Review for Environmental Strategies. vol. 6, no. 2, pp. 229–250. 2006.

[15] A. Muramatsu, T. Watanabe, A. Sasaki, H. Ito and A. Kajihara. Rice production with minimal irrigation and no nitrogen fertilizer by intensive use of treated municipal wastewater. Water Science and Technology. vol. 70, no. 3, pp. 510–516. 2014.

[16] T. P. Tuong and B. A. M. Bouman. Rice production in water-scarce environments. Water. vol. 5, pp. 53–67. 2003.

[17] A. K. Rath, B. Swain, B. Ramakrishnan, D. Panda, T. K. Adhya, V. R. Rao and N. Sethunathan. Influence of fertilizer management and water regime on methane emission from rice fields. Agriculture, Ecosystems and Environment. vol. 76, no. 2, pp. 99–107. 1999.

[18] Sustainable Rice Platform. Rice facts. [Online]. Available: http://www.sustainablerice.org/rice_facts.html. [Accessed 2016. 10. 30].

[19] J. McLean, D. Dawe, B. Hardy and G. Hettel. Rice Almanac: Third Edition. CABI Publishing. Oxford. England. 2002.

[20] C. G. Yoon, S. K. Kwun and J. H. Ham. Effects of treated sewage irrigation on paddy rice culture and its soil. Irrigation and Drainage. vol. 50, no. 3, pp. 227–236. 2001.

[21] M. S. Kang, S. M. Kim, S. W. Park, J. J. Lee and K. H. Yoo. Assessment of reclaimed wastewater irrigation impacts on water quality, soil, and rice cultivation in paddy fields. Journal of Environmental Science and Health, Part A: Toxic/Hazardous Substances and Environmental Engineering. vol. 42, no. 4, pp. 439–445. 2007.

[22] F. Papadopoulos, G. Parissopoulos, A. Papadopoulos, A. Zdragas, D. Ntanos, C. Prochaska and I. Metaxa. Assessment of reclaimed municipal wastewater application on rice cultivation. Environment Management. vol. 43, no. 1, pp. 135–143. 2009.

[23] S. Li, H. Li, X. Liang, Y. Chen, Z. Cao and Z. Xu. Rural wastewater irrigation and nitrogen removal by the paddy wetland system in the Tai Lake region of China. Journal of Soils Sediments. vol. 9, no. 5, pp. 433–442. 2009.

[24] T. I. Jang, H. K. Kim, C. H. Seong, E. J. Lee and S. W. Park. Assessing nutrient losses of reclaimed wastewater irrigation in paddy fields for sustainable agriculture. Agricultural Water Management. vol. 104, pp. 235–243. 2012.

[25] K. Jung, T. Jang, H. Jeong and S. Park. Assessment of growth and yield components of rice irrigated with reclaimed wastewater. Agricultural Water Management. vol. 138, pp. 17–25. 2014.

[26] A. M. Nyomora. Effect of treated domestic wastewater as source of irrigation water and nutrients on rice performance in Morogoro, Tanzania. Journal of Environment and Waste Management. vol. 2, no. 2, pp. 47–55. 2015.

[27] Q. W. Yang, C. Y. Lan, H. B. Wang, P. Zhuang and W. S. Shu. Cadmium in soil-rice system and health risk associated with the use of untreated mining wastewater for irrigation in Lechang, China. Agricultural Water Management. vol. 84, no. 1–2, pp. 147–152. 2006.

[28] D. T. Trang, W. van der Hoek, P. D. Cam, K. T. Vinh, N. Van Hoa and A. Dalsgaard. Low risk for helminth infection in wastewater-fed rice cultivation in Vietnam. Journal of Water and Health. vol. 4, no. 3, pp. 321–331. 2006.

[29] D. T. Trang, W. Van Der Hoek, N. D. Tuan, P. D. Cam, V. H. Viet, D. D. Luu, F. Konradsen and A. Dalsgaard. Skin disease among farmers using wastewater in rice cultivation in Nam Dinh, Vietnam. Tropical Medicine and International Health. vol. 12, no. Suppl. 2, pp. 51–58. 2007.

[30] H. P. Rhee, C. G. Yoon, Y. K. Son and J. H. Jang. Quantitative risk assessment for reclaimed wastewater irrigation on paddy rice field in Korea. Paddy and Water Environment. vol. 9, no. 2, pp. 183–191. 2011.

[31] Y.-J. An, C. G. Yoon, K.-W. Jung and J.-H. Ham. Estimating the microbial risk of *E. coli* in reclaimed wastewater irrigation on paddy field. Environmental Monitoring and Assessment. vol. 129, no. 1–3, pp. 53–60. 2007.

[32] V. Mukherjee and G. Gupta. Toxicity and profitability of rice cultivation under wastewater irrigation: the case of the East Calcutta Wetlands. South Asian Network for Development and Environmental Economics (SANDEE). vol. 62–11, pp. 292–300, 2011.

[33] Y. K. Son, C. G. Yoon, H. P. Rhee and S. J. Lee. A review on microbial and toxic risk analysis procedure for reclaimed wastewater irrigation on paddy rice field proposed for South Korea. Paddy and Water Environment. vol. 11, no. 1–4, pp. 543–550. 2013.

[34] Metcalf & Eddy, Inc. and AECOM Company, T. Asano, F. L. Burton, H. L. Leverenz, R. Tsuchihashi and G. Tchobanoglous. Water reuse: Issue, Technology, and Application. McGraw Hill, New York, USA. 1570pp. 2007.

[35] A. Sonune and R. Ghate. Developments in wastewater treatment methods. Desalination. vol. 167, no. 1–3, pp. 55–63. 2004.

[36] M. B. Pescod. Wastewater treatment and use in agriculture. FAO Irrigation and Drainage Paper. vol. 47, pp. 169. 1992.

[37] United States Environmental Protection Agency (EPA). Wastewater Treatment Works: The Basics. EPA 833-F-98-002. 1998.

[38] N. N. Thu. Urbanization and wastewater reuse in peri-urban areas: a case study in Thanh Tri District, Hanoi City. IWMI Work. Paper. no. 30, pp. 16–17. 2001.

[39] A. Muramatsu, H. Ito, A. Sasaki, A. Kajihara and T. Watanabe. Cultivation of rice for animal feed with circulated irrigation of treated municipal wastewater for enhanced nitrogen removal: comparison of cultivation systems feeding irrigation water upward and downward. Water Science and Technology. vol. 72, no. 4, pp. 579–584. 2015.

[40] D. Mara. Domestic wastewater treatment in developing countries. Earthscan, London, England. pp. 32. 2004.

[41] M. P. Ortega-Larroceaa, C. Siebe, G. Bécard, I. Méndez and R. Webster. Impact of a century of wastewater irrigation on the abundance of arbuscular mycorrhizal spores in the soil of the Mezquital Valley of Mexico. Applied Soil Ecology. vol. 16, no. 2, pp. 149–157. 2001.

[42] N. Rogan, T. Serafimovski, M. Dolenec, G. Tasev and T. Dolenec. Heavy metal contamination of paddy soils and rice (*Oryza sativa* L.) from Kocani Field (Macedonia). Environmental Geochemistry and Health. vol. 31, no. 4, pp. 439–451, 2009.

[43] WHO. WHO Guidelines for the Safe Use of Wastewater, Excreta and Greywater. Volume 2: Wastewater Use in Agriculture. vol. II, pp. 222. 2006.

[44] Stephen Foster, Héctor Garduño, Albert Tuinhof, Karin Kemper and Marcella Nanni. Sustainable groundwater groundwater management?: management urban wastewater as groundwater recharge evaluating and managing the risks and benefits. World Bank Briefing Note Series. vol. 12, pp. 6. 2005.

[45] H. Gallard and U. Von Gunten. Chlorination of natural organic matter: kinetics of chlorination and of THM formation. Water Research. vol. 36, no. 1, pp. 65–74. 2002.

[46] S. Toze. Reuse of effluent water—benefits and risks. Agricultural Water Management. vol. 80, no. 1–3 SPEC. ISS., pp. 147–159. 2006.

[47] A. Peasey, U. Blumenthal, D. Mara and P. G. Ruiz-palacios. A review of policy and standards for wastewater reuse in agriculture: A Latin American perspective. No. 68, Part II. Water and Environmental Health at London and Loughborough (WELL), London, England. 74p. June 2000.

[48] K. P. Singh, D. Mohan, S. Sinha and R. Dalwani. Impact assessment of treated/untreated wastewater toxicants discharged by sewage treatment plants on health, agricultural, and

environmental quality in the wastewater disposal area. Chemosphere. vol. 55, no. 2, pp. 227–255. 2004.

[49] K. Fytianos, G. Katsianis, P. Triantafyllou and G. Zachariadis. Accumulation of heavy metals in vegetables grown in an industrial area in relation to soil. Bulletin of Environmental Contamination and Toxicology. vol. 67, no. 3, pp. 423–430, 2001.

[50] J. O. Duruibe, M. O. C. Ogwuegbu and J. N. Egwurugwu. Heavy metal pollution and human biotoxic effects. International Journal of Physical Sciences. vol. 2, no. 5, pp. 112–118. 2007.

[51] A. Singh, R. K. Sharma, M. Agrawal and F. M. Marshall. Risk assessment of heavy metal toxicity through contaminated vegetables from waste water irrigated area of Varanasi, India. Food and Chemical Toxicology. vol. 51, no. 2 Suppl., pp. 375–387. 2010.

[52] IPPC. Mitigation of climate change: Contribution of working group III to the fourth assessment report of the Intergovernmental Panel on Climate Change. 2007.

[53] J. Zou, S. Liu, Y. Qin, G. Pan and D. Zhu. Sewage irrigation increased methane and nitrous oxide emissions from rice paddies in southeast China. Agriculture, Ecosystems and Environment. vol. 129, no. 4, pp. 516–522. 2009.

[54] Y. Jiao, Y. Huang, L. Zong, X. Zheng and R. L. Sass. Effects of copper concentration on methane emission from rice soils. Chemosphere. vol. 58, no. 2, pp. 185–193. 2005.

[55] M. A. Ali, J. H. Oh and P. J. Kim. Evaluation of silicate iron slag amendment on reducing methane emission from flood water rice farming. Agriculture, Ecosystems and Environment. vol. 128, no. 1–2, pp. 21–26. 2008.

[56] Y. Xu, J. Ge, S. Tian, S. Li, A. L. Nguy-Robertson, M. Zhan and C. Cao. Effects of water-saving irrigation practices and drought resistant rice variety on greenhouse gas emissions from a no-till paddy in the central lowlands of China. Science of Total Environment. vol. 505, pp. 1043–1052. 2015.

[57] T. Watanabe, T. Mashiko, R. Maftukhah and Nobuo Kaku, D. D. Pham and H. Ito. Rice cultivation and power generation circulated irrigation of treated municipal wastewater. Water Science and Technology in press.

[58] T. Watanabe, S. Kurashima, D. D. Pham, K. Horiguchi, T. Sasaki and J. Pu. Nutrient characteristics of rice for animal feed cultivated with continuous irrigation of treated municipal wastewater. Journal of Japan Society of Civil Engineers, Ser. G (Environmental Research), vol. 72(7), III_505-III_514 (in Japanese).

Chapter 4

On the Use of Decision-Support Tools for Improved Irrigation Management: AquaCrop-Based Applications

Joost Wellens, Dirk Raes and Bernard Tychon

Additional information is available at the end of the chapter

http://dx.doi.org/10.5772/109528

Abstract

Feeding more people with less water is putting efficient irrigation practices worldwide high on the agendas. As a reaction, over the last decades, numerous irrigation decision-support tools have been developed. For several reasons, the gap between farmer and modeler remained in most cases too large. The Food and Agriculture Organization of the United Nations (FAO) contributes to alleviate the encountered adoption limitations with AquaCrop and its stand-alone AquaCrop plug-in. This simple and robust field-crop-water balance has been successfully tested for a wide range of crops and regions, and its database is still expanding through worldwide contributions. The present chapter describes how AquaCrop can help irrigation advisory services draft efficient irrigation calendars that are easily applicable and adoptable: either by the elaboration of site-specific irrigation schedule calendars in chart format when the user has no access to the needed data or by the integration of its plug-in in a server/client ICT application offering centralized data management. As for the irrigation charts, studies prove 10-30% water savings, while maintaining yield and requiring minimum data. The server/client application offers an all-in advice tool, including real-time irrigation advice and yield forecasts. No adoption assessments have yet been carried out, but several ongoing pilot studies are promising.

Keywords: irrigation advisory service, decision-support tools, AquaCrop model, water-use efficiency, irrigation charts, ICT

1. Introduction

Irrigation is worldwide being considered as one of the means to increase or secure food production. As a result, in many parts of the world, the pressure on the available water resources has also intensified and is facing its limits. The challenge for the next decades will be how to feed ever more people with ever less water resources. The management of irrigation systems is, in most cases, based on the farmer's experience instead of rational basis (use of agro-climatic data) [1, 2]). The inefficiencies detected in the irrigation process [3, 4] have driven the development of tools to facilitate farmers the irrigation scheduling. Stakeholders need practical decision-support tools to help them assess irrigation practices and the resulting yield. Simulation models provide a low-cost means of investigating a wide range of management options.

Despite the plethora of irrigation scheduling support systems that have been developed over the past decades, there is little evidence of widespread adoption by farmers [2, 5]. Most producers find state of art irrigation scheduling tools overwhelming and lack the skills necessary to install, operate and troubleshoot them [1, 6]. Often not all required data are available at parcel level (real-time climate data and soil characteristics) or crops are not (yet) taken into account by the irrigation advice service [7]. Or the variables provided (e.g., daily crop water requirements) require additional calculations to transform it into useful data, namely management variables (e.g., daily irrigation time) [2]. Still another explanation for the low adoption rate is that farmers are not confident whether their use would actually transform into benefits [6, 8, 9]. This chapter presents some promising results and perspectives to bridge the gap between farmers and modelers, and overcome the above-stated limitations. The approach helps irrigation advisory services in the elaboration of efficient irrigation calendars that can be easily used by farmers, profiting hence to advisers and producers.

The Food and Agriculture Organization of the United Nations (FAO) has developed AquaCrop, a field-crop-water-productivity simulation model for use as a decision-support tool in planning and analysis [10, 11]. Being a water-driven crop model, crop biomass and harvestable yield are simulated in response to available water (soil moisture and irrigation). Although constructed upon basic and complex biophysical processes, only a relative small number of parameters are needed to adapt AquaCrop to different cases and crops. Often the integrated default input variables are sufficient and do not require additional fitting. When additional variables are needed, they are mostly intuitive and can easily be determined using simple methods [10, 12].

AquaCrop has been broadly tested for different corps around the world under diverse environments: for example, barley in sub-Saharan Africa [13], wheat in Iran [14] and in western Canada [15], teff in Ethiopia [16], quinoa in Bolivia [17] and maize in California [11]. Freely downloadable at FAO's website, the model contains a default database of the world's major crops (cotton, maize, potato, quinoa, rice, soybean, sugar beet, sunflower, tomato, wheat, barley, sugar cane, sorghum and teff [18]), and the list of crops is ever growing due to worldwide contributions. It has also been used to design different deficit irrigation strategies [19], to evaluate sowing strategies in a semiarid environment [20] and to develop an economic model for decision-support system at the farm scale [21]. An open source and animated Zotero Internet forum maintains an updated list of all peer-reviewed journal papers and Ph.D. manuscripts published on the calibrations and applications of the AquaCrop model [22].

How easily significant water savings can be obtained based on AquaCrop-derived simulation results is presented further down based on a case study in Burkina Faso. However, a quick literature review yields ample examples on the benefits of irrigation advice services. Water savings, while maintaining the same yield, can reach from 10 to 30% when compared with water use recorded in previous irrigation seasons without irrigation advice [1, 2, 8]. Other examples managed to combine yield increase and water-use reduction; for example, Eching [23] indicated an 8% yield increase with a 13% water-use reduction. The operating costs for an irrigation advisory service in Spain, including several field visits from technicians, are estimated at about 3 € ha^{-1} $year^{-1}$ [7]. Lorite et al. [25] studied the average annual irrigation benefits of shifting to irrigation advisory services ranging from 100 € ha^{-1} (for wheat and maize) to more than 400 € ha^{-1} (for sugar beet, sunflower and olive). Unfortunately, benefit assessments on irrigation advisory services are rare. The few existing financial studies confirm irrigation scheduling services are highly profitable [24] and encourage the integration of economic indicators in order to contribute toward a greater acceptance of advisory services [25].

The actual used irrigation advice methods and tools can be grouped into two approaches: the one using long-term averages and the other based on real-time data. The real-time approach requires access to daily weather data in conjunction with water budget calculators or crop models that rerun and update their output each time new data are available. The long-term averages approach is less complicated and does not require access to daily weather data (apart from rainfall, if applicable) [8]. For a given climate and crop, an irrigation calendar or different irrigation scenarios are elaborated only once and stuck throughout the growing season. AquaCrop and AquaCrop plug-in [26], a stand-alone executable deprived of its graphical user interface but offering all the same possibilities as the full program, offer the possibilities to play on both fronts.

2. Model description

FAO developed and freely distributes the AquaCrop model. This dynamic crop-growth model predicts yield response to water; it assumes a linear relationship between biomass (*B*) growth rate and crop transpiration (*Tr*). Only a small amount of water taken by the roots is used for growth and metabolism (i.e., biomass), the remaining is lost by transpiration. The transpiration rate depends directly on the available soil moisture; as more water depletes, *Tr* becomes less than potential, and biomass growth will reduce. For more details on the water and other stress mechanisms, refer also the studies of Steduto et al. [30]. In AquaCrop, actual crop transpiration is translated into biomass through a water productivity (*WP*) parameter. For a given time interval, the accumulated biomass is a result of the *WP* and the accumulated canopy transpiration: $B = WP \times \Sigma Tr$ [27–29]. The *WP* defines the amount of biomass a crop can produce per unit of water consumed. It is a crop-specific parameter. When normalized for atmospheric evaporative demand, $WP^* = B/\Sigma(Tr/ETo)$, it remains virtually constant over a range of environments [29] in order to make AquaCrop applicable across diverse locations and seasons. For most crops, only part of the biomass produced is partitioned to the harvested organs to give yield. The harvestable yield (*Y*) is portioned from the biomass by means of another crop-specific parameter known as the harvest index (*HI*): $Y = HI \times B$. The model does not include underlying hierarchical processes simulating the intermediary steps leading to biomass accumulation. As a consequence, the model structure is simple with few input parameters [10]. The model uses the more easily obtainable canopy ground cover (*CC*, the fractional coverage of the soil by the canopy) instead of leaf area index (*LAI*) as the basis for calculating transpiration and monitoring crop development.

AquaCrop uses two different kinds of parameters: (i) fixed or conservative parameters and (ii) case-specific or nonconservative parameters. Conservative parameters are independent of geographical region, management techniques or time. They should be determined under non-limiting growing conditions but remain valid for stress conditions through the integration of stress response functions [10, 12]. These conservative parameters consist mainly of canopy cover growth (*CGC*) and decline (*CDC*); crop coefficient for transpiration at full canopy (K_c); *WP* for biomass; and soil water depletion thresholds. These parameters are applicable to a wide range of different conditions and crop cultivars [30]. Some other crop parameters are case-specific and nonconservative (e.g., sowing density, length of phenological stages). Nonconservative parameters are affected by climate, field practices and soil conditions. The operator needs to provide them for each specific case and cannot apply them broadly. If not available, the model can offer estimations [12, 31]. FAO has calibrated crop parameters for several crops and provides them as default values in the crop files stored in the AquaCrop database. At the same time, literature on supplemental crops keeps growing.

New crops can be added to the database or an existing crop tested for a new agro-climatological region. In a first step, the calibration and validation procedure mainly bases on the monitoring of the fractional canopy cover evolution. **Figure 1** shows how the evolution of the fractional canopy cover can be followed by taking overhead pictures of the canopy throughout the growing season and quantified by means of image analysis software (e.g., [32]). Once the evolution of the fractional canopy cover is plotted (*CC*), the most important crop characteristics can be derived from this curve: days to emergence and maturity, canopy growth coefficient (*CGC*), canopy decline coefficient (*CDC*), etc. In a second step, *WP* and *HI* can be fine-tuned in order to match at best simulated versus observed yield.

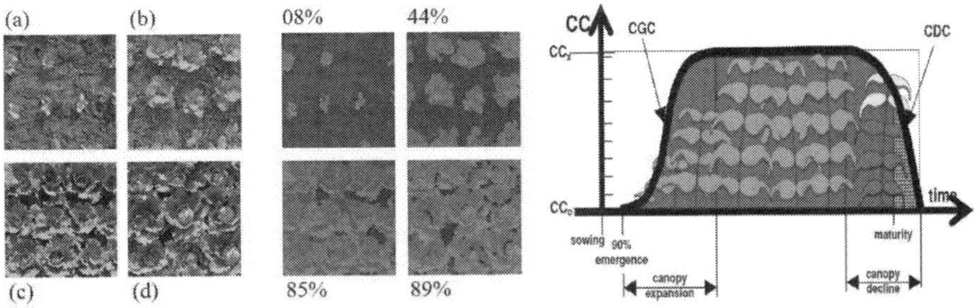

Figure 1. Left (a-d): overhead pictures of a cabbage field in Burkina Faso; middle: derived fractional cover data using image analysis software [32]. Right: canopy cover (CC) development curve and most important crop parameters [31] (CC_o: initial canopy cover; CC_x: maximum canopy cover; *CGC*: canopy growth coefficient; *CDC*: canopy decline coefficient).

For each day of the simulation period, AquaCrop requires minimum and maximum air temperature, reference evapotranspiration as a measure of the evaporative demand of the atmosphere and rainfall. Additionally, the mean annual CO_2 concentration has to be known (AquaCrop provides an historical time series of mean annual atmospheric CO_2 concentrations measured at Mauna Loa Observatory in Hawaii). The needed soil hydraulic characteristics, describing the soil water retention and soil water movement in the soil, are as follows: (i) the hydraulic conductivity at saturation: the ease with which water moves through a completely wetted soil; (ii) the soil water contents at saturation: the soil is completely filled with water and there is no air left; (iii) field capacity: the amount of soil moisture after excess water has drained away and (iv) permanent wilting point: the minimum soil moisture at which a plant wilts. One can either make use of the indicative values provided by the model for various soil texture classes or import locally determined data.

When all data are available (measured, estimated or adapted), AquaCrop offers the possibility to (i) determine net irrigation requirements in a given environment; (ii) assess an existing irrigation schedule and (iii) in the framework of the present chapter, to generate an irrigation schedule according to specified criteria. To generate an irrigation schedule for evaluating or planning a particular irrigation strategy, the irrigation method (sprinkler, drip or surface, which determines the fraction of the soil surface wetted by irrigation), and the time and depth criteria have to be specified. Time and depth criteria used for the generation of irrigation schedules are listed in **Table 1** [33].

	Parameter
Time criterion	
Fixed interval (days)	Time interval between irrigations (e.g., 10 days).
Allowable depletion (mm water)	Amount of water that depletes from the root zone (the reference is soil water content at field capacity) until irrigation is needed (e.g., 30 mm).
Allowable depletion (% of RAW)	Percentage of RAW that depletes until irrigation is needed (e.g., 100%).
Depth criterion	
Back to field capacity (± extra mm water)	Extra water on top of the required dose to bring the soil water content back to field capacity. Values can be zero, positive or negative.
Fixed application depth (mm water)	The irrigation amount that infiltrates in the field.
Water layer between bunds (mm water)	Threshold for the depth of the surface water layer that should be maintained between the soil bunds (e.g., 5 mm) for the generation of irrigation events for flooded rice.

Table 1. Types of time and depth criteria used for generating irrigation schedules [33].

3. Irrigation charts

Since crop water requirements vary over the growing season, farmers will need to adjust the irrigation during the season. Irrigation calendars are developed to give farmers simple guidelines on how to adjust their irrigation during the growing season. Site-specific calendar-based irrigation scheduling, accounting for local weather conditions and soil characteristics, provides irrigators with an inexpensive yet reliable strategy to estimate irrigation timing and amount [34]. In the design of these calendars, the irrigation water doses are usually considered as fixed. Fixed application depths in combination with variable irrigation intervals lead to the efficient use of irrigation water [35]. The selected value for the fixed application depth depends on the soil type, crop type, irrigation method and equipment. For the sake of simplicity and in order to promote adoption by farmers, the number of irrigation scheduling calendars is kept to a minimum, which means there has to be some generalization. The calendars for each crop are normally based on two planting dates, the major soils and perhaps two different initial soil water content values at the beginning of the irrigation season [36].

The procedure consists of two steps. To obtain reliable guidelines, the simulations need to be run for a long series of historical data. The historical data consist of daily air temperature, and daily, 10-day or monthly reference evapotranspiration. In general, no big variations are to be expected over the years, so one can directly work with mean values. Local soil physical characteristics and crop characteristics of the local variety need to be considered in the simulations. When done, the generated schedules with varying irrigation intervals during the different growth stages are simplified and translated into an easy readable chart.

3.1. Site-specific calendars

AquaCrop was assessed in several cabbage fields in Burkina Faso [32]. Few field data were required. Weather and soil data were provided by the responsible state agencies. Irrigation calendars were registered. Gravimetric soil water content was measured weekly at intervals of 0.2 m up to a depth of 0.6 m. These measurements were repeated three times per treatment, enabling the soil water balance simulation to be evaluated. All needed supplementary drop data were derived for each field by taking weekly dozens of overhead photos (2 m above the canopy) [37].

Figure 2 provides an example of simulated and observed soil water contents in a cabbage plot in Burkina Faso. Field monitoring started 3 weeks after planting. Soil water content exceeded field capacity during most of the growing season. When the soil water content is superior to field capacity, the excess water cannot be bound to the soil particles and drains, leading to water losses by percolation. AquaCrop was prepared to optimize irrigation schedules using local weather, soil and crop data (**Figure 3**). After a first irrigation for field preparation, initial soil water content was assumed to be at field capacity. In the area basin irrigation using a standard motor pump (±30 m^3/h) is the most common practice, and in general, gross application depths of 35 mm are applied. Deep percolation almost certainly occurs as it is nearly impossible to achieve uniform water distribution within a field and the correct rate of water application at the crop level. Since mostly surface irrigation, a field application efficiency of 0.6 was assumed [38]. The resulting soil water is given in **Figure 4**. Soil water content remained well below field capacity and above the readily available amount of water (*RAW*) threshold. *RAW* is the water that a plant can easily extract from the soil. When *RAW* is depleted from the root zone, the soil water content reaches a threshold at which the roots are no longer able to absorb sufficient water to the transpiration demand, and the plant is water-stressed. When soil water content reaches the permanent wilting point, no water remains available for the crops, and the crops will not be able to recover. By keeping the soil water content between field capacity and the *RAW* threshold, water losses due to deep percolation are limited, and crop water stress and yield loss are avoided. A reduction of ±20% in water use is registered, from 555 to 455 mm, while obtaining the same yield, ±52 ton/ha. 20% less water is hence being extracted from the river and becomes available for other crops and farmers, constituting a considerable benefit for the region as a whole.

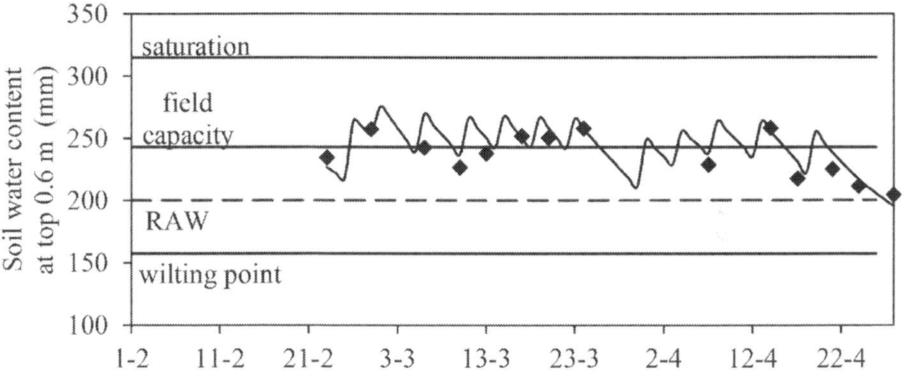

Figure 2. Observed (dots) vs. simulated (line) soil water content for a cabbage plot in Burkina Faso. Each dot is the average of three data. Irrigation: 555 mm; drainage:75 mm; yield: 52 ton/ha.

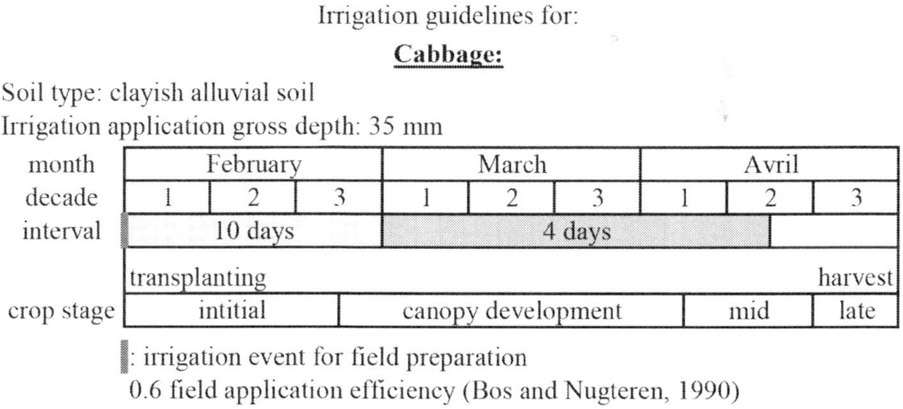

Figure 3. Example of an irrigation chart for cabbage. Cultivated in Burkina Faso on a clayish alluvial soil.

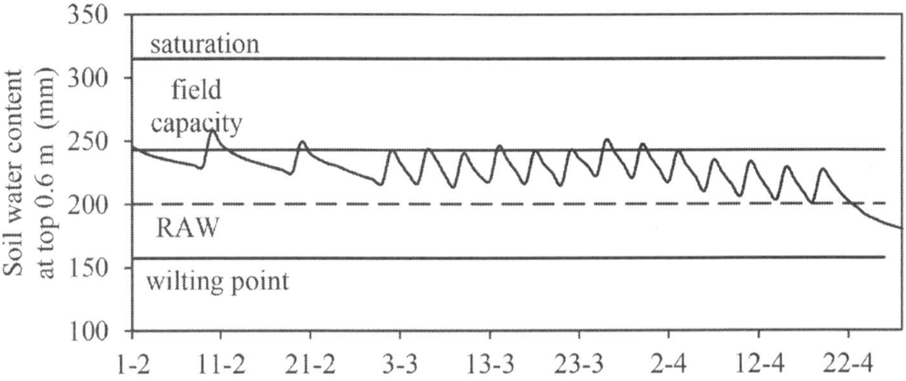

Figure 4. Soil water content when the proposed irrigation schedule is followed. Irrigation: 455 mm; drainage: 1 mm; yield: 53 ton/ha.

With the help of extension workers, the irrigation chart presented in **Figure 3** can be used by farmers [4]. Field application efficiencies, based on the findings of Bos and Nugteren [38], are already included in the gross irrigation application depths. On the back side, indicative durations in hours are given for different motor pump characteristics and field area. Different charts are being elaborated for the region's major crops, soils and irrigation systems.

3.2. Variations and accuracy

For a chart developed for a particular region, it is possible to assign standard weather conditions by analyzing the probability levels of rainfall in that area during different seasons. This could lead to icons to, for example, dry, wet or normal season next to an adapted irrigation calendar [1]. Raes et al. [35, 39] present slightly more elaborated charts for supplementary irrigation, when irrigation is combined with varying levels of rainfall. A variation for deficit irrigation is presented by Geerts et al. [19]. Water applications are limited to drought-sensitive growth stages, in order to maximize water productivity and stabilize, rather than maximize, yields.

To develop a simple tool factors needed to be simplified. In that process, some accuracy is lost. However, Boesveld et al. [1] found that simplified irrigation calendars exceeded detailed irrigation requirement based on modeling by only 2.7% and yielded water savings of 14%. In Fessehazion et al. [34], simple irrigation calendars gave similar irrigation applications, water losses and yields compared to real-time scheduling.

3.3. Real-time scheduling

Previous work has focused on scheduling irrigation over long time frames such as seasonal water allocations. Real-time irrigation scheduling, for example, hourly or daily, has received little attention [40]. Olivier and Singels [41] found that rainfall data and other observations by farmers were often unreliable and propose to go toward a centralized data processing and model execution. A centralized approach also counters the problems of preliminary training and collection of specific data which are not often available for most potential users [6, 42]. Actual research is developing smartphone applications to deliver these kinds of optimal irrigation calendars in near real time, based on daily weather data and information inputted by farmers through the application.

FAO also provides the AquaCrop plug-in program, performing identical calculation procedures as the AquaCrop standard program [26]. This version comes as a stand-alone executable without the user interface of the "classic" AquaCrop. By running the program, a list of projects, which contain all the required information for a simulation run, is carried out, and results are stored in output files. The plug-in program facilitates the inclusion of AquaCrop in external applications where iterative or large numbers of runs are required (e.g., [43]).

Figaro (*Flexible and precise irrigation platform to improve farm scale water production*, [44]), BELCAM (*Belgian collaborative agriculture monitoring*, [45]) and iPot (*industrial Potato monitoring*, [46]) are some recent examples of how AquaCrop plug-in is being integrated in information and communications technology (ICT) for agricultural advise, mainly focusing on irrigation and yield prediction. The platforms contain a database with crop, soil and real-time climate data. Farmers are invited to add the location of their fields and basic management characteristics (crop type, date of planting, etc.). A freely available and adapted Java script [47] picks up the required data, launches the AquaCrop plug-in executable and reads out the simulation results: up to date irrigation advice and yield forecast.

Figure 5 sketches the workflow of this approach for a case study in Belgium. A centralized database contains the soil map of Belgium, near real-time meteorological observations covering the whole of the country, a crop database with calibrated crop files for the major crops (potato, maize and winter wheat) and, if entered by farmers, information on field management practices (date of sowing, plant density, etc.), and otherwise default management values are used. On the server, the necessary input files are atomically generated, the AquaCrop plug-in executable launched and its output added to the central database. The main outputs for the moment are yield prevision and irrigation calendars. Through a web application, the data are available at parcel level and summarized for regional level, for the individual farmer but also for all concerned in the domains of agriculture (regional yield estimates and previsions) and environment (regional water balance). For the success of an advisory service, it is necessary to offer the information and tools, which are useful to the farmers and the society, as a whole (environmental benefit, food security, etc.) [7].

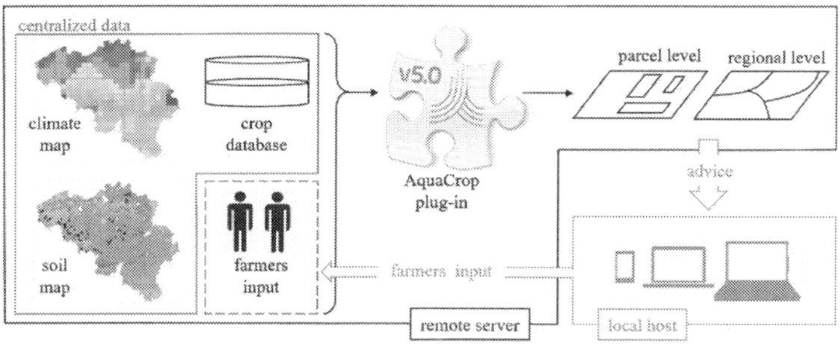

Figure 5. Real-time irrigation advice workflow (β-version under development for Belgian farmers).

4. Discussion and Conclusion

By 2030, irrigated land is predicted to increase by 28% [48], and the pressure on the available water resources will be considerable, even disastrous for some regions. Since a couple of decades, a myriad of irrigation decision-support tools has seen the light to help farmers obtain higher water-use efficiencies. However, a large gap still persists between available efficient water-use technologies and their adoption. Principal reason is the relatively little attention being paid to assist farmers in the adoption of new technologies. The models were often too complicated, high on input demands and/or too specific for only some crops. And local irrigation traditions were not taken into account, and the financial benefits were not clear [7].

FAO developed a simple and robust water-driven field crop model, AquaCrop and its standalone AquaCrop plug-in. The model, which comes already with a large crop database, requires a relatively small number of explicit and often intuitive data and does not require additional fitting. Once calibrated and validated, adapted irrigation schedules can easily be created. Based on AquaCrop simulations, irrigation calendar charts have been developed for use in Belgium, Tunisia, Mozambique and Burkina Faso. For the case of Burkina Faso, water savings amounted to ± 20% when using the proposed irrigation charts, while maintaining the same yield. When no real-time climate data are available, site-specific calendars may be more applicable. These simple and indicative irrigation charts are being transferred by extension workers in order to promote irrigation water savings and thus increase water availability for other users or crops. No data are available on the adoption rate of these irrigation charts. However, a survey was conducted to assess farmers' satisfaction with the overall irrigation advisory service. The general response was very positive, exceeding 90%. Twenty-one percent said they had seen an improvement in their livelihood because of better water distribution, thanks to water savings [49].

The same procedure has also been automated in a client/server application for agricultural advice (yield estimate and irrigation) in Belgium. Data management is centralized, and farmers can have access to personalized irrigation advice when logging into the website. Farmers are invited to add supplemental management information in order to improve the simulation results and the resulting advice; otherwise default, but locally correct, values are used. The tool is being developed in close collaboration with agricultural cooperation and technical centers so that farmers' expectations are taken into account. The application is still in its testing phase, so no information on farmers' appraisal and adoption is yet available. Once fully operational, the adoption rate could be easily evaluated by the numbers of farmers logged into the general server.

External factors and direct and indirect benefits will drive more and more farmers to subscribe to advisory services. Nowadays, the pressure exerted on the agricultural sector by public administration and clients to shift production forms a focus on quantity to a focus on sustainability, and quality is increasing worldwide [2]. The recent implementation of water pricing water policies, as already being required under the European Water Framework Directive, will motivate farmers to invest in technologies (such as decision-support irrigation tools) for improving water management [5]. In general, where resources are scarce and application costs are high, adherence to irrigation advice is also high [7]. Moreover, Qiao et al. [50] documented that farmers participating in an irrigation management program gained additional know-how that improved the water-use efficiency. Also, it is possible that those farmers involved in irrigation scheduling services more easily adopt other management recommendations leading to yield improvements [24].

Author details

Joost Wellens[1*], Dirk Raes[2] and Bernard Tychon[1]

*Address all correspondence to: joost.wellens@ulg.ac.be

1 Water, Environment & Development Unit, Environmental Sciences and Management Department, University of Liege, Liege, Belgium

2 Division of Soil & Water Management, Department of Earth & Environmental Sciences, K.U. Leuven University, Leuven, Belgium

References

[1] Boesveld, H., Zisengwe, L.S., Yakami, S., 2011. A simple irrigation scheduling tool for smallholder drip farmers. Irrigation and Drainage Systems, 25(4), 323–333. doi:10.1016/j.agwat.2006.12.011

[2] González Perea, R., Fernández García, I., Martin Arroyo, M., Rodríguez Díaz, J.A., Camacho Poyato, E., Montesinos, P., 2016 (in press, corrected proof). Multiplatform application for precision irrigation scheduling in strawberries. Agricultural Water Management. doi:10.1016/j.agwat.2016.07.017

[3] García Morillo, J., Rodríguez Díaz, J.A., Carnacho, E., Montesinos, P., 2015. Linking water footprint accounting with irrigation management in high value crops. Journal of Cleaner Production, 87, 594–602. doi:10.1016/j.jclepro.2014.09.043

[4] Wellens, J., Traore, F., Diallo, M., Tychon, B., 2013. A framework for the use of decision-support tools at various spatial scales for the management of irrigated agriculture in West-Africa. Agricultural Sciences, 4 (8A), 9–15. doi:10.4236/as.2013.48A002

[5] Giannakis, E., Bruggeman, A., Djuma, H., Kozyra, J., Hammer, J., 2016. Water pricing and irrigation across Europe: opportunities and constraints for adopting irrigation scheduling decision support systems. Water Sciences & Technology: Water Supply, 16(1), 245–252. doi:10.2166/ws.2015.136

[6] Mannini, P., Genovesi, R., Letterio, T., 2013. IRRINET: large scale DSS application for on-farm irrigation scheduling. Procedia Environmental Sciences, 19, 823–829. doi:10.1016/j.proenv.2013.06.091

[7] Ortega, J.F., de Juan, J.A., Tarjuelo, J.M., 2005. Improving water management: the irrigation advisory service of Castilla-La Mancha (Spain). Agricultural Water Management, 77, 37–58. doi:10.1016/j.agwat.2004.09.028

[8] Singels, A., Smith, T., 2006. Provision of irrigation scheduling advice to small-scale sugarcane farmers using a web-based crop model and cellular technology: a South African case study. Irrigation and Drainage, 55, 363–372. doi:10.1002/ird.231

[9] Morrison, M., 2009. Encouraging the adoption of decision support systems by irrigators. Rural Society, 19, 17–31. doi:10.5172/rsj.351.19.1.17

[10] Steduto, P. Hsiao, T.C., Raes, D., Fereres, E., 2009. AquaCrop—The FAO crop model to simulate yield response to water. I. Concepts and underlying principles. Journal of Agronomy, 101, 426–437. doi:10.2134/agronj2008.0139s

[11] Hsiao, T.C., Heng, L., Steduto, P., Rojas-Lara, B., Raes, D., Fereres, E., 2009. AquaCrop – The FAO crop model to simulate yield response of water: III. Parameterization and testing for Maize. Agronomy Journal, 101 (3), 448–459. doi:10.2134/agronj2008.0218s

[12] Raes, D., Steduto, P., Hsiao, T.C., Fereres, E., 2009. AquaCrop – The FAO crop model to simulate yield response to water. II. Main algorithms and software description. Journal of Agronomy, 101, 438–447. doi:10.2134/agronj2008.0140s

[13] Araya, A., Habtu, S., Hadgu, K.M., Kebede, A., Dejene, T., 2010. Test of AquaCrop model in simulating biomass and yield of water deficient and irrigated barley (Hordeum vulgare). Agricultural Water Management, 97, 1838–1846. doi:10.1016/j.agwat.2010.06.021

[14] Andarzian, B., Bannayan, M., Steduto, P., Mazraeh, H., Barati, M.E., Barati, M.A., Rahnama, A., 2011. Validation and testing of the AquaCrop model under full and deficit irrigated wheat production in Iran. Agricultural Water Management, 100, 1–8. doi:10.1016/j.agwat.2011.08.023

[15] Mhkabela, M.S., Bullock, P.R., 2012. Performance of the FAO AquaCrop model for wheat grain yield and soil moisture simulation in Western Canada. Agricultural Water Management, 110, 16–24. doi:10.1016/j.agwat.2012.03.009

[16] Araya, A., Keesstra, S.D., Stroosnijder, L., 2010. Simulating yield response to water of teff (Eragrostis tef) with FAO's AquaCrop model. Field Crops Research, 116, 196–204. doi:10.1016/j.fcr.2009.12.010

[17] Geerts, S., Raes, D., Garcia, M., Miranda, R., Cusicanqui, J.A., Taboada, C., Mendoza, J., Huanca, R., Mamani, A., Condori, O., Mamani, J., Morales, B., Osco, V., Steduto, P., 2009. Simulating yield response of quinoa to water availability with AquaCrop. Agronomy Journal, 101 (3), 499–508. doi:10.2134/agronj2008.0137s

[18] Raes, D., Steduto, P., Hsiao, T.C., Fereres, E., 2011. AquaCrop—Reference Manual—Annexes. FAO, Rome, Italy, 69 p. http://www.fao.org/nr/water/aquacrop.html

[19] Geerts, S., Raes, D., Garcia, M., 2010. Using AquaCrop to derive deficit irrigation schedules. Agricultural Water Management, 98, 213–216. doi:10.1016/j.agwat.2010.07.003

[20] Tsegay, A., Vanuytrecht, E., Abrha, B., Deckers, J., Gebrehiwot, K., Raes, D., 2015. Sowing and irrigatin strategies for improving rainfed tef (Eragrositis tef (Zucc.) Trotter) production in the water scarce Tigray region, Ethiopia. Agricultural Water Management, 150, 81–01. doi:10.1016/j.agwat.2014.11.014

[21] García-Vila, M., Fereres, E., 2012. Combining the simulation crop model AquaCrop with an economic model for the optimization of irrigation management at farm level. European Journal of Agronomy, 36(1), 21–31 doi:10.1016/j.eja.2011.08.003

[22] Zotero, 2016. AquaCrop Publications. https://www.zotero.org/groups/aquacrop_publications

[23] Eching, S., 2002. Role of technology in irrigation advisory services: the CIMIS experience. In: Proceeding of workshop on irrigation advisory services and participatory extension in irrigation management. FAO-ICID, Montreal, 1–12. http://www.fao.org/nr/water/docs/ias/paper24.pdf

[24] Montoro, A., López-Fuster, P., Fereres, E., 2011. Improving on-farm water management through an irrigation scheduling service. Irrigation Science, 29, 311–319. doi:10.1007/s00271-010-0235-3

[25] Lorite, I.J., García-Vila, M., Carmona, M.-A., Santos, C., Soriano, M.-A., 2012. Assessment of the Irrigation Advisory Services' recommendations and farmers' irrigation management: a case study in Southern Spain. Water Resources Management, 26, 2397–2419. doi:10.1007/s11269-012-0023-3

[26] Raes, D., Steduto, P., Hsiao, T.C., Fereres, E., 2012. Reference Manual: AquaCrop Plug-in Program (Version 4.0). FAO, Rome. http://www.fao.org/nr/water/aquacrop.html

[27] de Wit, C.T., 1958. Transpiration and crop yields. Versl. Landbouwk. Onderz. 64.6 Institute of Biological Chemistry Research On Field Crops and Herbage, Wageningen, The Netherlands. 90 p. http://edepot.wur.nl/186445

[28] Hsiao, T.C., Bradford, K.J., 1983. Physiological consequences of cellular water deficits. In: Taylor, H.M., Jordan, W.A., Sinclair, T.R., eds. Limitations to Efficient Water Use in Crop Production. Madison, Wisconsin, USA, ASA, pp. 227–265. doi:10.2134/1983.limitationstoefficientwateruse

[29] Steduto, P.E., T.C. Hsiao and E. Fereres, 2007. On the conservative behavior of biomass water productivity. Irrigation Science, 25(3), 189–207. doi:10.1007/s00271-007-0064-1

[30] Steduto, P., Hsiao, T.C., Fereres, E. Raes, D., 2012. Crop yield response to water. FAO Irrigation and Drainage Paper N° 66. Rome, Italy. 500 p. http://www.fao.org/docrep/016/i2800e/i2800e00.htm

[31] Raes, D., Steduto, P., Hsiao, T.C., Fereres, D., 2011. AquaCrop – The FAO Crop Model to Simulate Yield Response to Water: Reference Manual. www.fao.org/nr/water/aquacrop.html

[32] Wellens, J., Raes, D., Traore, F., Denis, A., Djaby, B., 2013. Performance assessment of the FAO AquaCrop model for irrigated cabbage on farmer plots in a semi-arid environment. Agricultural Water Management, 127, 40–47. doi:10.1016/j.agwat.2013.05.012

[33] Raes, D., Van Gaelen, H., 2015. Book II – Running AquaCrop. AquaCrop Training Handbooks. FAO, Rome, Italy. 122 p.

[34] Fessehazion, M.K., Annadale, J.G., Everson, C.S., Stirzaker, R.J., van der Laan, M., Truter, W.F., Abraha, A.B., 2014. Performance of simple irrigation scheduling calendars based on average weather data for annual ryegrass. African Journal of Range & Forage Science, 31(3), 221–228. doi:10.2989/10220119.2014.906504

[35] Raes, D., Sahli, A., Van Looij, J., Ben Mechlia, N., Persoons, E., 2000. Charts for guiding irrigation in real time. Irrigation and Drainage Systems, 14, 343–352. doi:10.1023/A:1006412031535

[36] Hill, R.W., Allen, R.G., 1996. Simple irrigation calendars: a foundation for water management. In: Food and Agricultural Organization of the United Nations (FAO) (Ed.), Irrigation Scheduling: From Theory to Practice. Rome, Italy. pp. 69–74. http://www.fao.org/docrep/w4367e/w4367e00.htm

[37] Hu, Z, F. He, J, Yin, X. Lu, S. Tang, L. Wang and X. Li, 2007. Estimation of fractional vegetation cover based on digital camera survey data and a remote sensing model. Journal China University of Mining and Technology, 17 (1), 116–120. doi:10.1016/S1006-1266(07)60025-X

[38] Bos, M.G., Nugteren, J., 1990. On irrigation efficiencies. ILRI Publication 19. Wageningen, The Netherlands. 120 p. http://library.wur.nl/WebQuery/wurpubs/fulltext/71061

[39] Raes, D., Smith, M., De Nys, E., Holvoet, K., Makarau, A., 2002. Charts with indicative irrigation intervals for various weather conditions. In: Proceeding of workshop on irrigation advisory services and participatory extension in irrigation management. FAO-ICID, Montreal, 1–11. http://www.fao.org/nr/water/docs/ias/paper3.pdf

[40] Saleem, S.K., Delgoda, D.K., Ooi, S.K., Dassanayake, K.B., Kiu, L., Halmaguge, M.N., Malano, H., 2013. Model predictive control for real-time irrigation scheduling. 4th IFAC Conference on Modelling and Control in Agriculture, Horticulture and Post Harvest Industry. August 27–30, Espoo, Finland. 299–304. doi:10.3182/20130828-2-SF-3019.00062

[41] Olivier, F.O., Singels, A., 2004. A survey of irrigation scheduling practices in the South African sugar industry. Proceedings of the South African Sugar Technology Associations, 78, 239–244. http://www.sasta.co.za/publications/congress-proceedings

[42] Domínguez, A., Martínez, R.S., de Juan, J.A., Martínez-Romero, A., Tarjuelo, J.M., 2012. Simulation of maize crop behavior under deficit irrigation using MOPECO model in a semi-arid environment. Agricultural Water Management, 107, 42–53. doi:10.1016/j.agwat.2012.01.006

[43] Lorite, I.J., García-Vila, M., Santos, C., Ruiz-Ramos, M., Fereres, E., 2013. AquaData and AquaGIS: Two computer utilities for temporal and spatial simulations of water-limited yield with AquaCrop. Computers and Electronics in Agriculture, 96, 227–237. doi:10.1016/j.compag.2013.05.010

[44] FIGARO Irrigation Platform, 2016. http://www.figaro-irrigation.net/

[45] BELCAM – Belgian Collaborative Agriculture Monitoring, 2016. http://maps.elie.ucl.ac.be/belcam/

[46] Wellens, J., Bernard, T., Piccard, I., Gobin, A., Curnel, Y., Goffart, J.P., Planchon, V., Cattoor, N., Cools, R., 2015. iPot: industrial Potato monitoring for the Belgian potato sector using remote sensing and crop growth modelling. World Potato Congress, 28–30 July, Beijing, China. http://hdl.handle.net/2268/184503

[47] GitHub, 2016. Interface for a local or remote AquaCrop instance. https://github.com/itszootime/aquacrop-interface

[48] FAO, 2002. World agriculture: towards 2015/2030 – Summary report. FAO, Rome, Italy. 106 p. http://www.fao.org/3/a-y3557e.pdf

[49] Wellens, J., Nitcheu, M., Traore, F., Tychon, B., 2013. A public-private partnership experience in the management of an irrigation scheme using decision-support tools in Burkina Faso. Agricultural Water Management, 116, 1–11. doi:10.1016/j.agwat.2012.09.013

[50] Qiao, G., Zhao, L., Klein, K.K., 2009. Water use associations in inner Mongolia: factors that influence farmers to join. Agricultural Water Management, 40, 333–339. doi:10.1016/j.agwat.2008.11.001

Chapter 5

Technical Efficiency of the Subsurface Drainage on Agricultural Lands in the Moldova River Meadow

Oprea Radu, Sorin Mihai Cimpeanu,
Razvan Ionut Teodorescu and Daniel Bucur

Additional information is available at the end of the chapter

http://dx.doi.org/10.5772/67258

Abstract

This study aims to investigate the technical efficiency of different subsurface drainage variants, in terms of the depth of the tile drains, spacing between the drain lines, type and thickness of the drain + filter complex, and the improvement procedures. Within the four variants, the discharge rate of the soil moisture excess was studied. In variants A and D, the spacing between drains is 20 m, and in the variants B and E, the spacing is 15 m. The depth of the tile drains is 0.8 m in variants D and E and 1.0 m in variants A and B. In variant A, tile drainage was combined with land shaping in the bedding system with top of ridges and furrows. Soil moisture was determined on checkpoints placed on drain cross section, at 2 m from drain lines, and of the middle of the drain spacing. In the version with land shaping, the drain lines located under the furrows favor the excess moisture removal. A similar technical efficiency was recorded in unimproved variant but with spacing between drains of 15 m. Best efficiency at removing excess water was registered in variant of the filtering material from ballast associated with flax strains.

Keywords: moisture excess, tile drainage, land shaping, technical efficiency

1. Introduction

Soil quality is more or less affected by one or more conditions, such as drought, periodic excess water, soil erosion, landslides, etc. [1, 2]. Their harmful influences are reflected by the degradation of soil characteristics and functions, of soil bio-productive capacity, affecting yields and food safety, with consequences on human life quality [3–5]. These conditions are determined either by natural factors [6–8] or by agricultural and industrial activities that might have a negative synergic action [9, 10].

Some of the main limiting factors of crop production in Suceava County are excess water, floods, low infiltration rate and soil compaction, soil erosion, and landslides. Waterlogging is a complex process determined by water supply, retention, movement, and discharge in the sub-adjacent rock-soil system [11, 12]. Water excess (rainfall water and/or groundwater) gets manifest under different forms and intensities, both in case of flat and slope lands [13–15]. Natural conditions of the Baia Piedmont Plain favor the occurrence and maintenance of waterlogging. The Moldova River meadow and terraces platform in the form of stripes with an average width of 1.5 km, almost parallel to Moldova River bed, oriented from North West to South East, of gentle slope (1–5%), with flat-horizontal areas and a lot of micro-depressions, facilitating water storage. Because of the humid climate in the Radaseni-Fantana Mare-Baia depression, at low evapotranspiration, the 1–5 days of consecutive heavy rainfall are the main source of water excess in the poorly permeable soils. Furthermore, the hydrographic network represented by the Somuzul Baii runlet, which crosses the mid-area of Radaseni-Fantana Mare-Baia, with a riverbed of about 0.3–0.6 m depth and shallow at some sectors, is a permanent cause of excessive humidity, prone to flooding nearby agricultural lands during heavy rains. Groundwater is stored in permeable deposits of sands with gravel and rocks, and its level is generally free, yet in meadow and low terrace areas, at highest levels, it flows under pressure, causing rising of the water table.

Experimental drainage fields were organized to explore the wide range of causes conducive to excessive water, the complexity of natural conditions hosting the phenomenon, and the multitude of its forms of manifestation. Study results are the base for solutions applied in implementing planning projects of the said areas and, by extrapolation, they were the basic materials for spreading technical solutions in natural homogenous conditions, concretized in the elaboration of design and use standards, instructions, or recommendations.

2. Material and method

Based on the pedo-climatic conditions of the wet area of Suceava County (**Figure 1**) and from the humid area of the meadow and terrace platform of the Moldova, respectively, subsurface drainage experimental fields were organized as a main mean of rainfall excess removal, locally associated to different land improvement works.

The pilot experimental agricultural drainage field of Baia was organized in 1978 by the National Society for Land Improvements—Suceava branch—in collaboration with the Agronomical University of Iasi, on a pseudo-gley glosic albic luvisol, used in natural pastures and hayfields with periodic excessive humidity from precipitations.

The hydro-technical scheme of the Baia experimental drainage field (**Figure 2**) comprises a surface of 3.00 ha, organized by subdivided lots, with two repetitions of three variants each, distinguished by distance between drain lines (12; 15; 20 m), average depth of the pipe drains is 0.8–1.0 m, type and diameter of drain pipes, type and thickness of filtering materials (**Table 1**).

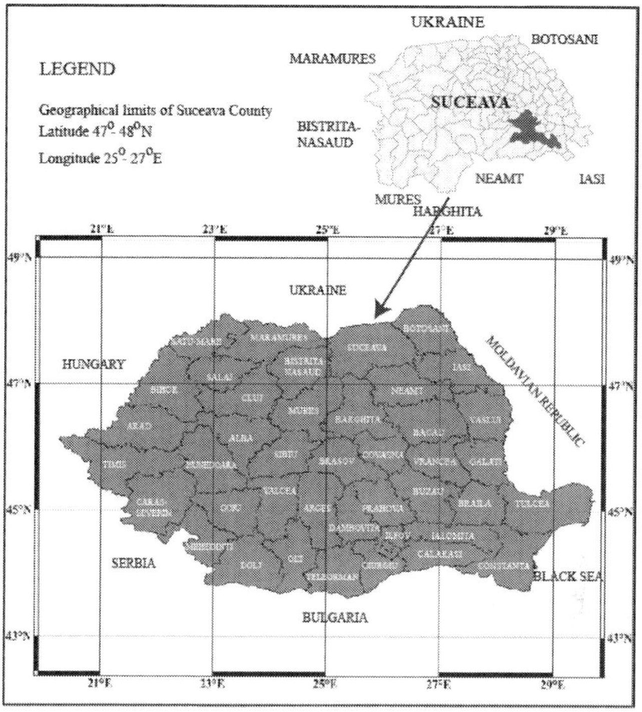

Figure 1. Map of Romania and the geographical position of the Baia drainage field.

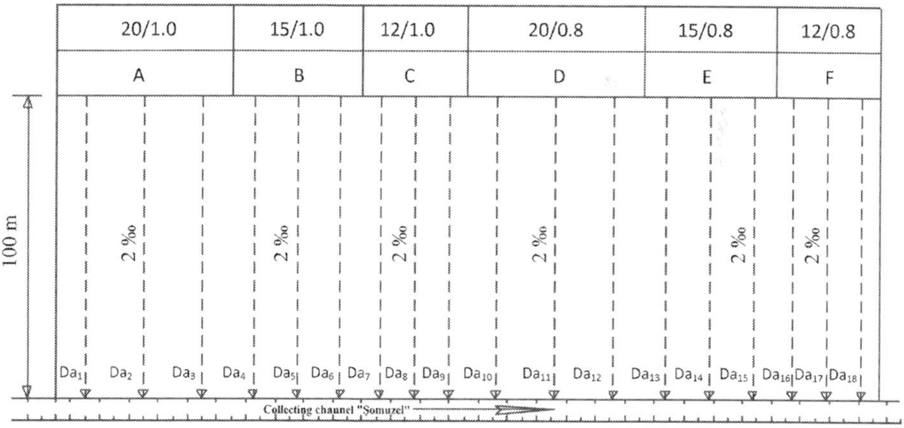

Figure 2. Baia experimental drainage field.

Tile drainage variant	Spacing between drain lines/depth drain (m)	Drain line number	Pipe type and diameter (mm)	Type and thickness of the drain + filter complex (cm)
A	20/1.0	1	Tile Ø 70	Ballast (20) + flax stems (50)
		2	Tile Ø 125	Ballast (70) + green sods
		3*	Tile Ø 70	Ballast (20) + green sods
B	15/1.0	4	Tile Ø 70	Flax strains (30)
		5*	Tile Ø 70	Ballast (12) + flax stems (20)
		6	Tile Ø 70	Ballast (15) + green sods
C	12/1.0	7	Corrugated plastic Ø 65	Ballast (12) + green sods
		8	Smooth plastic Ø 63	Ballast (12) + green sods
		9	Tile Ø 70	Ballast (15) + green sods
D	20/0.8	10	Corrugated plastic Ø 65	Ballast (20) + flax stems (40)
		11	Smooth plastic Ø 110	Ballast (60) + green sods
		12*	Tile Ø 70	Ballast (20) + green sods
E	15/0.8	13*	Tile Ø 70	Flax strains (30)
		14*	Tile Ø 70	Ballast (12) + flax stems (20)
		15*	Tile Ø 70	Ballast (15) + green sods
F	12/0.8	16	Corrugated plastic Ø 65	Ballast (12) + green sods
		17	Smooth plastic Ø 63	Ballast (12) + green sods
		18	Tile Ø 70	Ballast (15) + green sods

*Studied drain lines/variants.

Table 1. Constructive elements of drains from the Baia experimental drainage field.

To improve the production capacities of agricultural lands and mostly of farm fields, surface and subsurface drainage system, banking, watercourse regulation, soil erosion control, and other such works were conducted. According to data from the National Association for Land Improvement, in Suceava County, there is an area of 44,904 ha with surface drainage works, of which 27,455 ha were completed with underground drainage works. The drainage ditches are of 1875 km in length, while the subsurface drainage network has a total length of 11,909 km.

Based on results from the Baia agricultural drainages experimental field, technical solutions for the Moldova River meadow depletion-drainage planning were established. Rotopanesti-Radaseni-Fantana Mare, Dragoiesti-Berchisesti, Bogdanesti-Baia, and Baisesti-Dumbrava surface and subsurface drainage systems comprise a total surface of 8761 ha, of which underground drainage covered an area of 3059 ha (**Figure 3**).

The surface drainage ditch network with a total length of 126.85 km is made of collecting, disposal and interception channels, and so on. To discharge excessive soil moisture, an underground drainage network was organized, made of tile drains with a total length of 1575.12 km, based on the nature and intensity of excessive water.

The Baia agricultural experimental drainage field is placed in the northern part of the Rotopanesti-Radaseni-Fantana Mare depletion-drainage system, on a platform of the upper terrace, on the left side of the Moldova River, located North West from the Baia depression. In the natural conditions of this wetland, the albic stagnant-lossic luvisol is used as a natural pasture.

http://dx.doi.org/10.5772/67258

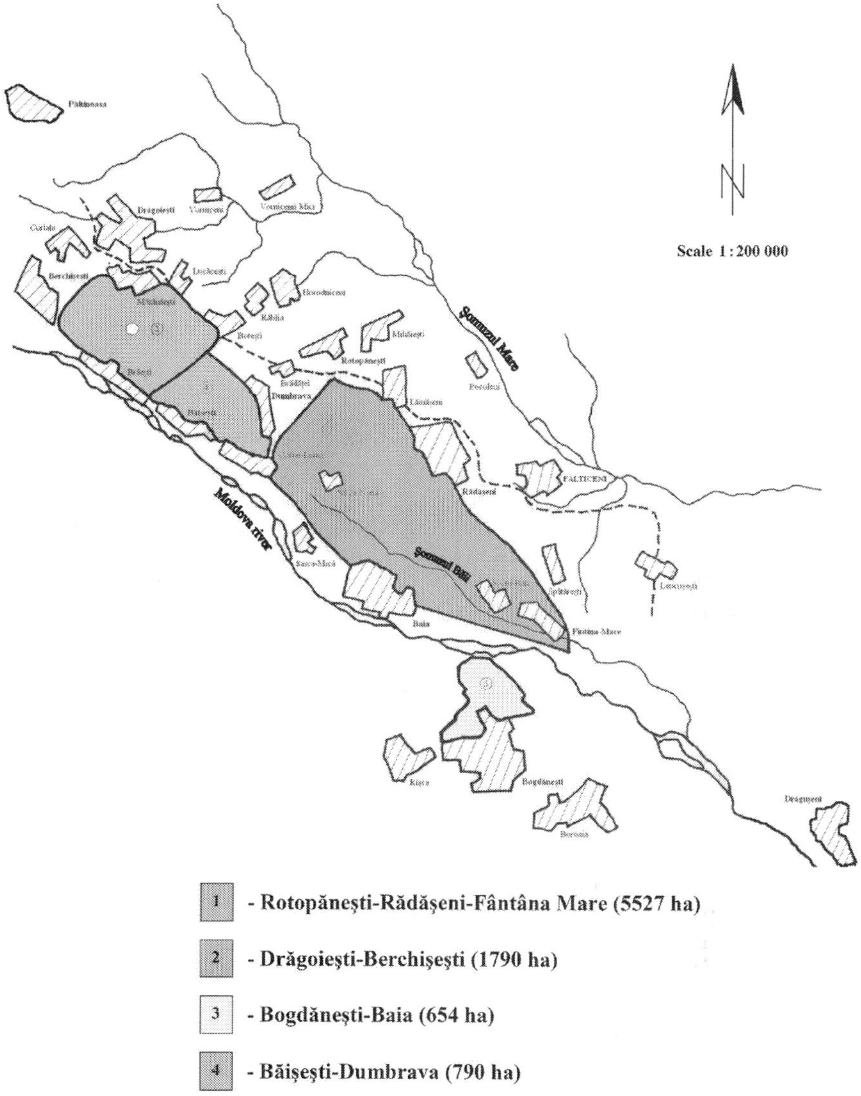

Figure 3. Surface and subsurface drainage systems of Moldova River meadow.

The surface and subsurface drainage systems, as a measure of moisture control, were associated to works of land shaping in the bedding system (**Figure 4**), deep soil loosening, mole drainage, amendments, and so on. After implementing improvement procedures, their operation and behavior are of high importance.

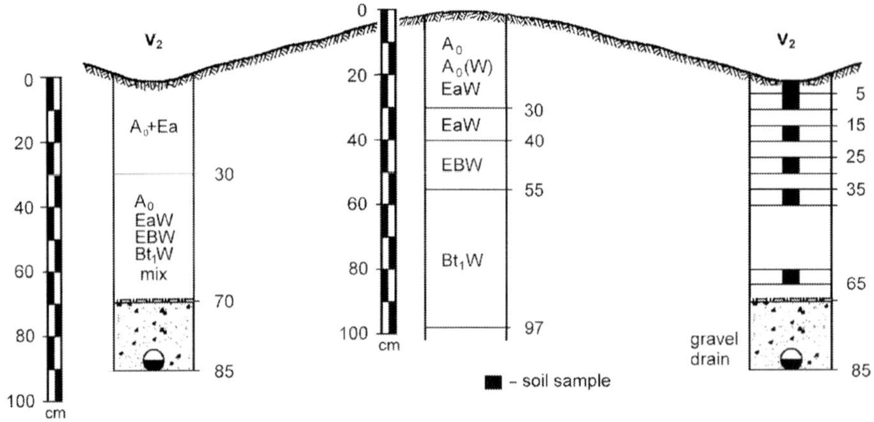

Figure 4. Tile drainage combined with the bedding system variant: top of ridges and furrow.

In order to evaluate the operative behavior of drainage technical solutions, after 36 years soil samples were collected to establish soil moisture in layers with thickness of 10 cm each, up to a depth of 0.8 m and 1.0 m, depending on the depth of the tile drains, at 2 and 5 days, respectively, after the rainfalls. Control points were placed on the drain line, 2.00 m from it and mid-distance between drain lines.

Soil samples were collected from the area served by D_5, D_{13}, D_{14}, and D_{15} drains, placed 15 m apart and laid at 0.8 m (D_{13}, D_{14}, D_{15}) and, respectively, laid at 1.0 m depth from D_5 drain and from the area served by D_3 and D_{12} drains which are 20 m apart, but laid at different depths, 0.8 m the D_{12} drain and 1.0 m the D_3 drain, with a ridge-and-furrow land surface.

3. Results and discussions

The analysis of soil moisture values at 10 cm depth, 2 days after cumulative rainfall of 40 mm, on the variants with the spacing between drains of 15 m, shows that in case of control points placed 2.0 m from the drain line and at mid-distance between drains, the moisture level got to the depth of 30–40 cm, where the poorer permeable layer is located and decreased (**Figure 5**). As for D_3 tile (distance between drain lines 20 m and depth of the pipe drains 1.0 m), with the land shaping in the bedding system with top of ridges and furrows, the soil moisture at control points located 2.0 m from the drain line and mid-distance between drains (10.0 m) got to the depth of 40–50 cm, due to the harsher material placed when the ridge-and-furrow system was created, which implicitly increased the depth of the poorer permeable layer (**Figure 6**). In all cases, at the control point located on the drain line, soil moisture values 2 days after rainfall generally increase as depth increases, as the water flow is directed to the drain line and the permeability of the filtering layer got poorer over the 36-year operating period.

http://dx.doi.org/10.5772/67258

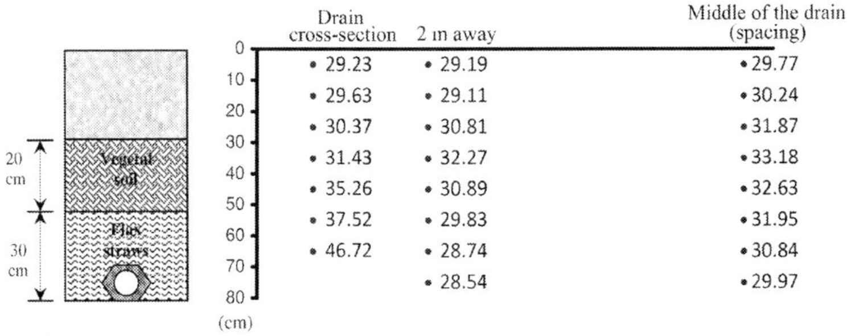

Figure 5. Soil water content in the area served by the tile drain D_{13}, 2 days after rainfall.

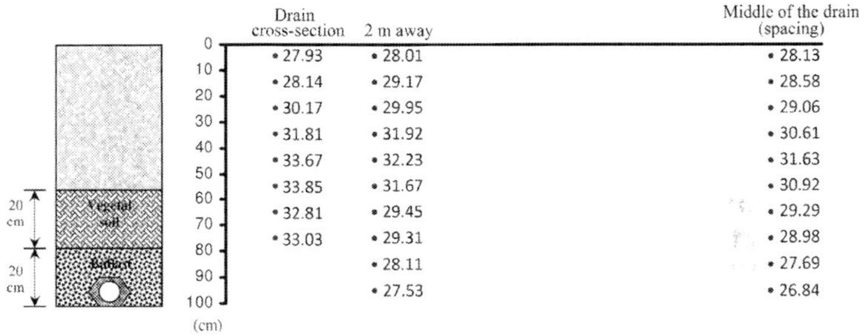

Figure 6. Soil water content in the area served by the tile drain D_3, 2 days after rainfall.

After 5 days from rainfall, soil moisture at control points of D_5, D_{12}, D_{13}, D_{14}, and D_{15} drains located at 2.0 m from the drain line and the middle of the drain spacing generally increases from a 10–20 cm depth to a 40–50 cm depth; then it slightly decreases because of the poorer permeable layer (**Figure 7**). Higher moisture in the superficial soil layer is due to water storage by organic material from the discontinuous celery layer, as the drained surface is used as pasture.

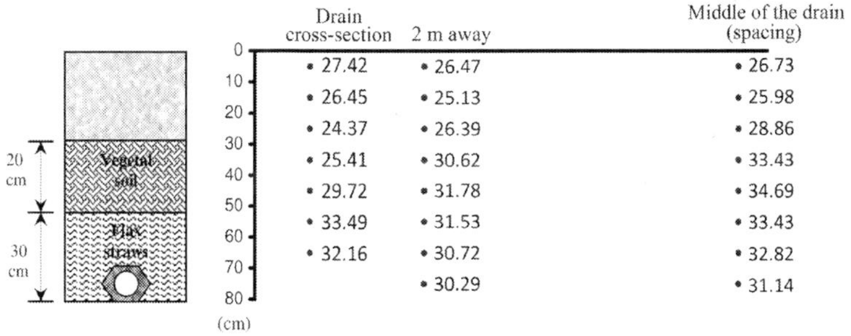

Figure 7. Soil moisture in the area served by the tile drain D_{13}, 5 days after rainfall.

At D_3 drain with ridge-and-furrow system, soil moisture values have the same dynamics; the only difference being that the depth where the greatest moisture is recorded increases from 50 to 60 cm for the control point located at 2.0 m from the drainage ditch to 60–70 cm mid-distance between drains, due to water excess elimination and water consumption by plants (**Figure 8**).

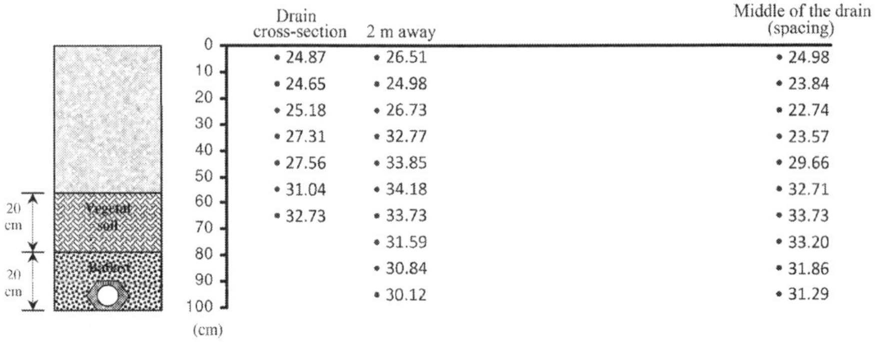

Figure 8. Soil water content in the area served by the tile drain D_3, 5 days after rainfall.

At the drain line, soil moisture values generally decrease to the depth of 20–30 cm, and then they increase thanks to the water flow directed to the drain filter during the operating period. For the first 20 cm, moisture values are higher because of the plant root system, which is better developed on the drain cross section.

The analysis of the values of the average content of soil water by control points, established 3 days from rainfalls, shows that at the drains placed at 15 m one from another, the lowest value is recorded at the control point located 2.0 m from the drain line, while the highest value is recorded on the drainage cross section (**Figure 9**).

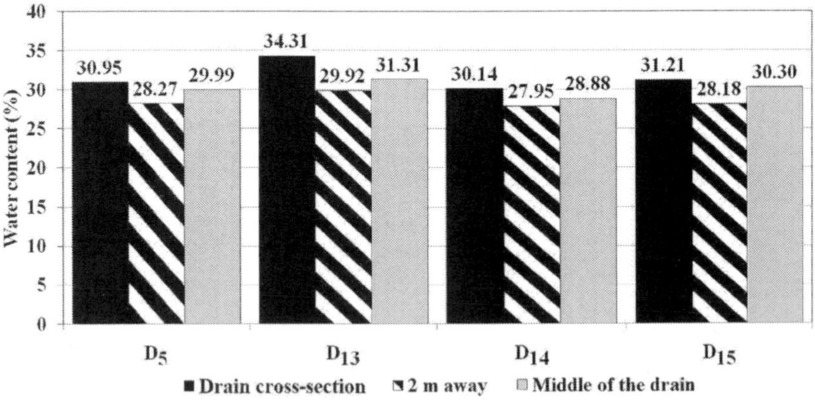

Figure 9. Average soil moisture at control points, after 2 days, in variant with spacing between drains of 15 m.

At D_3 drain with distance between drain lines of 20 m and with ridges and furrow, the highest value is also registered on the drainage ditch, but the lowest value was obtained at the control point located mid-distance between drain lines, due to the water flow directed to the drain filter and runoffs toward the drain line produced during heavy rainfalls (**Figure 10**).

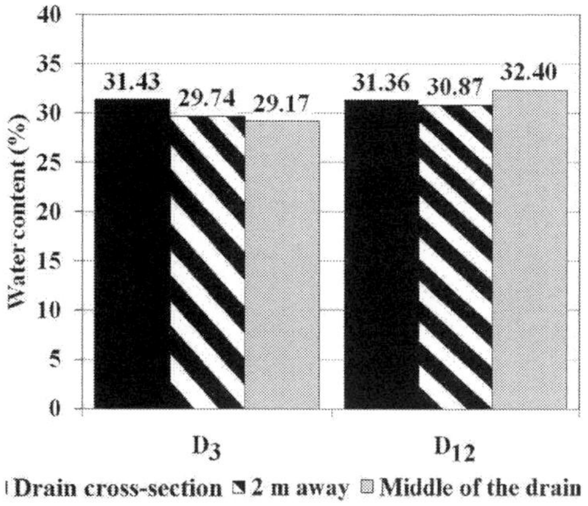

Figure 10. Average soil moisture at control points, after 2 days, in variant with spacing between drains of 20 m.

Runoff control toward the drain line in case of D_3 drain with a ridge-and-furrow land surface is also shown in the analysis of the average soil moisture recorded at D_{12} drain, with the distance between drain lines of 20 m, but with the non-modeled land surface, where the highest value was obtained for the control point located mid-distance between drains (**Figure 10**).

After 5 days, the last precipitations, the average moisture reaches the highest peak at the control point located mid-distance between drain lines, while the lowest value is registered for the drain cross section, except for D_3 drain (**Figures 11** and **12**). The decreasing values of average soil moisture water at mid-distance between drains to the drain line show drain system operation after 36 years of activity.

For the D_3 drain with ridges and furrow, the lowest value of the average moisture is also recorded at the drain line, yet the value obtained at the control point mid-distance between drains (on ridge) is 1.77% lower than the value at the control point 2.0 m away from the drain line, due to any runoff control to the drain line and to having added harsher material when the ridges were created. For such material, hydro-physical indices (fading coefficient, field water capacity) are lower than those of finer textured material (with high clay content).

The analysis of soil moisture content in the control section, 2 days and 5 days from the rainfall recording, of the studied drains (**Figure 13**), shows that the efficiency of water excess removal may be assessed by using different drainage technical solutions after 36 years of operation.

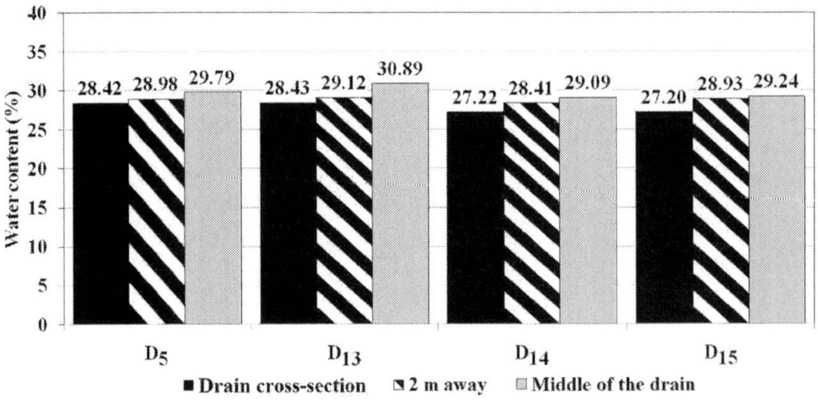

Figure 11. Average soil moisture at control points, after 5 days, in variant with spacing between drains of 15 m.

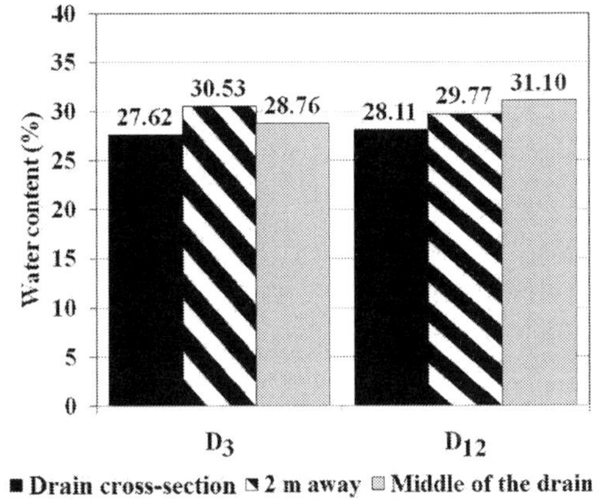

Figure 12. Average soil moisture at control points, after 5 days, in variant with spacing between drains of 20 m.

For D_{13}, D_{14}, and D_{15} tile drains laid at the same depth (0.8 m) and distance between drain lines of 15 m and with different filtering materials, the lowest average moisture values on the control section, in both soil sampling stages, were obtained for D_{14} drain. The use of a ballast layer of 12 cm thick and of a flax stalk layer of 20 cm as filtering material at D_{14} drain favored the discharge of excessive humidity for the serviced area, obtaining an average moisture of 28.99% every 2 days, respectively, and 28.24% every 5 days from rainfall recording, values which are by 0.91%, respectively, and 0.22% lower than those obtained at D_{15} drain, where the only ballast layer used was 15 cm thick.

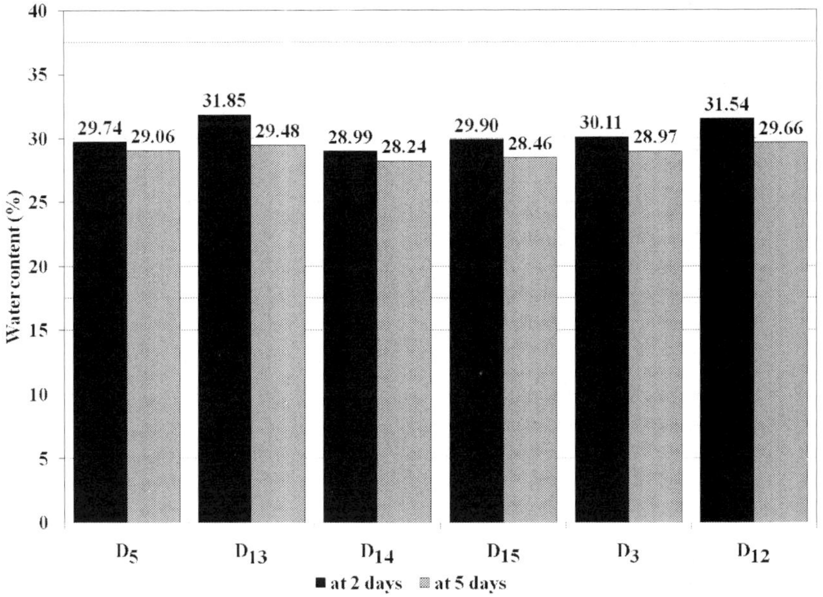

Figure 13. Average soil moisture on control section, at 2 and 5 days, respectively, after the rainfalls.

At D_{13} drain with a filtering layer made of 30 cm thick flax stalks, after 36 years of operation, they turned into organic matter, prone to poorer permeability around the drain pipe and water flow to the drain, recording 2 days from rain the highest average soil moisture at the control section of 31.85% and 5 days after the value of 29.48%. At the same drain, 2 days after the last rainfall, the greatest amplitude of the soil moisture of 4.39% per control points was achieved.

Maintaining the distance of 15 m between drains, the same filtering material (ballast layer of 12 cm thick and flax stalk layer of 20 cm thick) and increasing the laying depth from 0.8 m (D_{14}) to 1.0 m as for D_5 drain, there is an increase of average soil moisture for the control section, from 28.99 to 29.74% (29.94% calculated at 0.8 m depth) during the first soil sample collecting stage. Five days after rain recording, the average soil moisture for control section of D_5 drain increased by 0.82% (0.94% calculated at 0.8 m depth) in comparison with soil moisture determined by D_{14} drain.

At D_{12} drain laid at 0.8 m depth, when the distance between drain lines is of 20 m, high soil moisture values are recorded. After 2 days of rainfall, average soil moisture is 31.54%, by 2.55% higher than D_{14} drain, where the lowest value was recorded. During the second stage of sample collection, this drain registered the highest value of 29.66%, 1.42% higher than the value obtained in D_{14} drain.

By increasing the laying depth to 1.0 m, maintaining the distance of 20 m between drain lines and land shaping in the bedding system with top of ridges and furrow as in D_3 drain variant, a better discharge of excessive water is achieved, yielding an average soil moisture of 30.11%, respectively, and of 28.97% for the control section, values that are relatively close to those determined in case of the drains placed every 15 m, except the D_{13} drain.

4. Conclusions

- In case of subsurface drainage systems from the Moldova River meadow, the predominant spacing between drains is 20 m, and the depth of the tile drains is 1.0 m in the terrace area.
- For an efficient water excess removal, it is advisable improvement procedure as land shaping in the bedding system with ridges and furrows in order to ensure a good drainage.
- Ridge-and-furrow system favors the discharge of water excess, as the runoffs are led to drain lines, obtaining a better water collection and discharge during the first hours and days from excessive water incidence.
- The average soil moisture content of 2 days following the rainfall has the smallest value close to the drain lines, due to the water afflux toward the drains and of the decrease of the permeability of the filtering layer, in time.
- The use of the flax stalks only as filtering material, regardless of the thickness of the layer, is not recommended because of permeability diminishing. However, it was observed that their association with ballast ensures the best discharge of the moisture excess even after 35 years of operation.

Author details

Oprea Radu[1], Sorin Mihai Cimpeanu[2], Razvan Ionut Teodorescu[2] and Daniel Bucur[1*]

*Address all correspondence to: dbucur@uaiasi.ro

1 Faculty of Agriculture, University of Agricultural Sciences and Veterinary Medicine, Iasi, Romania

2 Faculty of Land Reclamation and Environmental Engineering, University of Agronomical Sciences and Veterinary Medicine, Bucharest, Romania

References

[1] Burja C., Burja V., 2013, Sustainable development of rural areas: a challenge for Romania, *Environmental Engineering and Management Journal*, 14(8), 861–1871.

[2] Hornbuckle J.W., Christen E.W., Faulkner R.D., 2007, Evaluating a multi-level subsurface drainage system for improved drainage water quality, *Agricultural Water Management*, 89(3), 208–216, http://dx.doi.org/10.1016/j.agwat.2007.01.004

[3] Mirsal A.I., 2004. Soil Pollution. Origin, Monitoring and Remediation. Springer: Berlin.

[4] Walker W.R., 1989. Guidelines for Designing and Evaluating Surface Irrigation Systems. Irrigation and Drainage Paper 45. FAO Rome, Italy.

[5] Valipour M., 2014. Comparison of different drainage systems usable for solution of environmental crises. In Soil. Proc. 1st International Conf. on Environmental Crises and Its Solutions, Kish Island, Iran, 8 pages.

[6] Troeh F.R., Thomson L.M., 2005. Soils and Soil Fertility. Blackwell Publisher: Iowa.

[7] Lukianas A., Vaikasas S., Malisauskas A., 2006, Water management tasks in the summer polders of the Nemunas Lowland, *Irrigation and Drainage*, 55(2), 145–156. John Wiley & Sons. DOI: 10.1002/ird.230.

[8] Lin D.G., Hung S.H., Ku C.Y., Chan H.C., 2016, Evaluating the efficiency of subsurface drainages for Li-Shan landslide in Taiwan, *Natural Hazards and Earth System Science*, Copernicus Publications, Göttingen, Germany, 1–22, DOI: 10.5194/nhess-2015–309.

[9] Irwin R.W., 1997. Handbook of Drainage Principles. Publication 73, RP-01-97-500. Ontario Ministry of Agriculture, Food and Rural Affairs (*OMAFRA*), Toronto, Canada.

[10] Townend J., Reeve M.J., Carter A., 2001. Water release characteristics. In Soil and Environmental Analysis: Physical Methods, (2nd edn), Smith K.A., Mullins C.E. (eds). Marcel Dekker, Inc.: New York.

[11] Kovács G., 1981. Seepage Hydraulics. Developments in Water Science vol. 10. Elsevier Science Publishers: Amsterdam.

[12] Singh V., Frevert D., Rieker J., Leverson V., Meyer S., Meyer S., 2006, Hydrologic modeling inventory: cooperative research effort, *Journal of Irrigation and Drainage Engineering*, 132(2), 98–103. DOI: 10.1061/ASCE

[13] Valipour M., 2013. Scrutiny of inflow to the drains applicable for improvement of soil environmental conditions. In Proc. 1st Int. Conf. on Environmental Crises and Its Solutions, At Kish Island, Iran, 12 pages.

[14] Tang Z., Shi C., Bi K., 2013, Impacts of land cover change and socioeconomic development on ecosystem service values, *Environmental Engineering and Management Journal*, 14(10), 2697–2705.

[15] Ritzema H.P., Kselik R.A.L., Chanduvi F., 1996. Drainage of Irrigated Lands: a Manual. Irrigation Water Management Training Manual. FAO: Rome.

Chapter 6

Applications of SuDS Techniques in Harvesting Stormwater for Landscape Irrigation Purposes: Issues and Considerations

Andrew B. Shuttleworth, Ernest O. Nnadi,
Fredrick U. Mbanaso, Stephen J. Coupe,
Joris G.W.F. Voeten and Alan P. Newman

Additional information is available at the end of the chapter

http://dx.doi.org/10.5772/67041

Abstract

While urbanization and increasing population has put much pressure on natural drainage channels and resulted in increase in flooding, there is increased pressure on available water resources due to climate change, reduction in frequency of rainfall events and drought. The emergence of a sustainable drainage system (SuDS), also known as best management practice (BMP) and low impact development (LID), has changed the management strategy of drainage from conventional to sustainable. SuDS techniques seek to deliver the three cardinal paradigms of sustainable drainage: quantity, quality and amenity and as such, they can offer an additional benefit for applications such as landscape irrigation. Most SuDS techniques have the potential for water storage with minimal or no modifications required. This chapter, while covering the capabilities of SuDS systems, explores SuDS devices such as pervious pavements equipped with excess storage capacity, cisterns and tanks harvesting roofwater, infiltration systems aimed at supporting the growth of urban plants and green roofs with the potential to store water in order to maintain water demanding planting scheme even during dry periods. It also covers systems where SuDS is the main driver to device installation and address issues and considerations surrounding applications of such systems in water harvesting for irrigation.

Keywords: Irrigation, Drainage, SuDS, Stormwater, rain harvesting

1. Introduction

Urbanization can initiate undesirable local modifications to the water cycle. In particular, the effects of the spread of impermeable area both within and at the margins of towns and cities have highlighted the need to modify the way that society deals with stormwater. The problems of surface flooding in cities and overloading of foul sewers (when combined systems are in place) are the most headline-grabbing aspects of increasing impermeability but reduction in aquifer recharge can also be important when water demand rises due to increase in local population.

In many western societies, particularly in the United Kingdom, there is a preference for housing schemes with gardens (including, often, very water demanding planting regimes). This increases the demands for water well beyond traditional domestic use. Alongside these domestic developments, hotels, hospitals, office blocks, light industrial units and retail developments often try to show a supposedly green face to the world in the form of extensive landscaped areas around their equally extensive impermeable parking surfaces. We are thus faced with the dual problem of a demand for water for landscaping accompanied by rapid and wasteful run off from impervious parking areas. This can cause flooding problems in response to short heavy rain events. The problem is particularly important in summer months when the demand for watering domestic, commercial and municipal planted areas is at its highest, at the same time as the supply from surface sources is at a minimum.

The term SuDS was originally used in the United Kingdom as an acronym for 'sustainable urban drainage systems' but this term has more recently lost favour and is commonly replaced by SuDS (sustainable drainage systems). In the USA, the terms low impact development and best management practices (LID and BMP) cover the same approach to drainage and in Australia the term 'water-sensitive urban design' is favoured. While all of the above terms are used, sometimes interchangeably, to indicate a holistic approach to stormwater management, they are all dominated in their philosophy by the concept of stormwater source control. This entails controlling both the quality and quantity of stormwater as close to its site of deposition as possible. Commonly, parts of the United Kingdom can be under drought orders (which impose summer hosepipe bans which limit landscape irrigation) and still suffer from localized flooding during short-term storms. These occurrences are becoming an increasing feature in some parts of the United Kingdom and this adds extra incentive to retrofitting of SuDS systems that can also make a contribution to the reduction of water use. The principles of SuDS are often presented as follows:

1. Storing runoff so that it can be released slowly or used beneficially (attenuation).
2. Allowing water to soak into the ground so as to mimic the processes on an undeveloped surface (infiltration).
3. Where necessary to allow the water to move doing so at the surface and at a controlled velocity.
4. Using the processes of collection, storage and transportation to facilitate the removal of pollutants by sedimentation, precipitation, adsorption (in its widest sense) and degradation.

Most SuDS devices have the potential to be used for water storage with minimal or no modifications required. **Table 1** summarises the irrigation water resource potential of a range of hard SuDS devices that have been used for water harvesting, whereas **Table 2** outlines a number of green SuDS devices that themselves place a potential demand on water resources but which can be modified to harvest water for self-irrigation and sometimes for other uses.

Device	Quantity issues	Quality issues
Underground and above ground barrels, tanks and cisterns harvesting roof water only	The sizing of the system needs to be done carefully and the usage of the water be well established if both stormwater attenuation and water harvesting are to be achieved in the same system. With modern computer control the sizing can be reduced by actively draining tanks in response to predicted rain events.	The quality of water will be affected mainly by atmospheric fallout and the nature of materials used in roof construction. While this is not really considered an issue new UK guidance requires that even roof water should have some treatment before discharge to a watercourse
Underground and above ground storage harvesting both surface runoff and roof water	The additional resource will require a greater storage volume for it to be utilisable. If there is attenuation available upstream of the storage tanks this can be minimized as a problem	Surface water will generally be of much poorer quality than roof water alone (see above) but provided the system is correctly designed to retain day to day pollutant releases it will generally only be an issue if major pollutants releases overcome the pollution attenuation mechanisms. Salt applications in temperate zones can be a problem.
Pervious pavements and similar without off line additional storage	Unless taking runoff from impervious surfaces too the total inputs are limited to the water falling on the surface. Unless designed with extra storage capacity even an attenuation based system will have limited capacity to store water for more than short periods of time	Input of potentially harmful organisms from faecal contamination and chemical pollutants from atmospheric fallout. If exposed to traffic there will also be day to day input of automotive based pollutants and if not provided with upstream protection the water can be subject to contamination from the very rare losses of engine oil and fuel. In temperate areas de-icing salts can be an issue.
Pervious pavements with off- line additional storage	By directing water into off line underground or above ground tanks the storage required for attenuation or to overcome limited rates of infiltration will not be compromised by the need to recover water	

Table 1. Irrigation water resource potential of a range of hard SuDS devices.

This chapter, while covering the capabilities of all SuDS systems, gives particular emphasis to those that are particularly suitable for retrofitting. In this chapter, the greatest attention is given to devices such as pervious pavements equipped with excess storage capacity, cisterns and tanks harvesting roofwater, infiltration systems deliberately aimed at improving the growth of urban plants (including trees) and green roofs, which can offer the potential to store water in excess of the capacity of the substrate so as to maintain a water demanding planting scheme during dry periods. Inevitably, there will be some overlap with harvesting for purposes other than irrigation. Competing use for harvested water includes toilet flushing and washing machine use and with appropriate treatment, many of the types of use that normally demand potable water. We must also recognize that rainwater harvesting systems are commonly provided without involvement of the SuDS philosophy, being designed without any attempt to enhance the control of stormwater (but sometimes doing so by accident). This chapter only covers systems where SuDS is the main driver to device installation.

Device	Quantity issues	Quality issues
Green roofs with additional off line storage Green roofs with integrated storage.	Both need careful sizing taking into account predicted available rainfall and demand placed by both growing plants and any additional off-roof demands. Storage needs to take into account sufficient temporary storage to provide stormwater source control when substrate is close to saturated. Provision may need to be made for rapid dumping of stored water when large storms are predicted in a 'reservoir full' situation.	If used for self-irrigation of the roof or for additional off-roof irrigation there will be little problem but if, as in some installations, green roof drainage water is pooled with other water and used for such as toilet flushing the colour can become a source of user complaint. If fertiliser is used on the roof, then recycling the water via the roof will retain the nutrients where they are needed.
Infiltrating urban tree planters with subsidiary methods of storage.	Maintaining sufficient moisture for the trees can be a problem. Proprietary systems which incorporate water storing foams (as used by florists) can offer a solution. Species selection for the variability of available water needs to be done carefully.	There will be day-to-day input of automotive based pollutants and if not provided with upstream protection the water can be subject to contamination from the very rare losses of engine oil and fuel. In temperate areas de-icing salts can be an issue.

Table 2. Green SuDS devices that can be modified to harvest water for self-irrigation and other uses.

Harvested rainwater from urban environments (even water directly from roofs) will inevitably contain pollutants that would not be expected if the water was collected from rural upland catchments or extracted from protected aquifers. If water is to be used for irrigation after being in contact with pollutant materials inevitably generated by both the daily activities of urban living and by fallout from industry, a first question must be whether the quality of the water is good enough for purpose. Irrigation water limits vary from place to place and not all organisations include all the possible pollutants (see Ref. [1]). A useful comparative table is provided by Nnadi et al. [2] which presents the irrigation water limits provided by a range of authorities.

2. Roofwater harvesting: rain barrels and cisterns

While many SUDS systems have been used to harvest rainwater for a combination of both irrigation and indoor uses (such as toilet flushing and washing machine pre-wash) it is the direct harvesting of roof water (and sometimes water from non-trafficked paved areas) into barrels and cisterns that has seen the greatest uptake as a component of the LID programmes in the USA. The relatively high quality of this source of harvest (although well

http://dx.doi.org/10.5772/67041

short of potable water standards in most jurisdictions (see Ref. [3]) that makes this source attractive for internal use as well as for irrigation. The primary differences between a cistern or rain barrel which forms part of a SuDS system and one which is simply a water supply device are the way the container is sized and the way the water is managed. Whether managed manually by an individual householder (possibly backed up by an advice and education program) or through an automated system, unless efforts are made to maintain enough available storage to contain at least a significant proportion of a design storm and unless the stored water is managed either by using the water or by releasing a proportion before the next storm then this is not SuDS at all but purely a water harvesting exercise. Jones and Hunt [4] made the point that the greatest problem with roofwater harvesting as a stormwater control mechanism is that many of the cisterns and barrels remain full all of the time because water is not utilized. Managing the cistern by using the water regularly or releasing during a dry period is necessary if roofwater storage is to be seen as SuDS.

North Carolina State University (NCSU) offers online a very comprehensive review of rainwater harvesting [5] and also one of the best sources of information provided for home owners on water cisterns and their use [4]. Their document usefully attempts to guide the owners in relation to the need for additional ground support for cisterns in gardens, an often overlooked factor. NCSU also provides an informative guide to roof water quality [6]. Many other American universities provide a similar extension literature and the various States of the USA represent such a broad diversity of climates that the guidance from either state regulators or that provided through local university extension programs should be selectable if someone is looking for information that suits their climate elsewhere in the world. That is not to say that authors from outside the USA have not been active in reporting their work in a variety of jurisdictions (although in many cases, the stormwater source control element is missing).

Cisterns can be either above ground or below ground, but the land take and unattractive appearance of reasonably sized water cisterns are often a barrier to adoption. Underground tanks can be expensive and offer significant structural problems if they are to be covered with a trafficked surface or even one with a reasonable dead load. Unlike above ground storage, where irrigation use can often be achieved under gravity flow, they will normally need a pump.

Underground tanks are not normally suitable for occupier retrofit even by the most capable householders. An option that addresses this is the Skeletank®, where the thin polyethylene tanks get their structural strength from internal interlocking polypropylene skeleton. This provides structural strength while maintaining a system that can be carried, if necessary, on the roof of a car and can be lifted easily by two people. They are easily linked together to achieve the size required. The system was originally designed to provide stormwater attenuation and soakaway tanks (used upside down) within individual domestic curtilages. The system has since been adopted for rainwater harvesting and lightweight pump and filter chambers are available as accessories. Apart from the wiring for the pumps, the system is totally suitable for installation by an amateur. **Figure 1** shows an installation in Preston (Lancashire, UK) which provides an interesting example of a householder built retrofit water harvesting SuDS project particularly since the primary aims of harvesting the water was to provide irrigation water to landscaped areas and to support a small pond on site while allowing total disconnection of

Figure 1. Left: Skeletanks® in place and connected together. Right: lightweight pump chamber.

drainage from the combined sewer. The owners of the property in question were undergoing conversion works which included the installation of a new patio area. The tanks were installed underneath the proposed patio area. The concept of the system begins with the collection of rainwater in a downpipe chamber. On this site, three downpipe chambers were installed, two chambers for a 94 m^2 roof area at the main house and the final chamber positioned to collect water from a 55 m^2 area (serving an adjacent outbuilding). These chambers are intended to pre-clean the rainwater and allow silt to deposit within the sump of the chamber. Water passes through a fabric filter on the outlet spigot of the chamber before running into the tanks which were connected together using standard underground drainage pipe to provide a combined reservoir. For this site, sixteen tank units, providing a total storage capacity of 4800 litres, were used. The required number of units was determined by a modelling to fit available storage to the various demands including sufficient temporary storage, with controlled discharge to accommodate stormwater during rain events when the temporary storage is full. The rainwater pump chamber is the final component in the chain and is primarily used to pump rainwater on demand for irrigating flower beds and grassed areas.

When available in excess stored rainwater is also pumped into a header tank for internal household use, such as, flushing toilets and water feed for washing machine pre-wash. The primary re-use of the harvested rainwater is to irrigate the flower beds and grassed areas surrounding the property and also for supporting the flora and fauna of the pond. Since the installation of the system, both use of mains water for irrigation and for topping up the pond has been minimal. Checks have shown the public health microbiological qualities (including legionella) are not a problem.

3. Pervious and macro-pervious pavements as a water harvesting system

Pervious pavement systems (PPS) have been around as SuDS elements since the 1970s and are extremely well documented (see Refs. [7–12]). In the United Kingdom, their origin is largely associated with Chris Pratt's design which utilized concrete block pavers provided with infiltration channels as the wearing course [13].

The pervious wearing course surface can also be porous asphalt or porous poured concrete. The wearing course is usually laid on a bedding layer of, typically, 10 mm material and there is often a geotextile as a separation and filtration layer between the bedding layers and either the subbase of aggregate (typically a uniform 50 mm, with a void ratio of about 30%) or on a sub-base replacement of plastic load-bearing boxes (a typical void ratio of 90%). If the pavement is to be used for water harvesting, it requires a reliable under-sealing layer which can be a folded or welded polyolefin membrane or, particularly within individual curtilages, can be conveniently provided by one or more of the sub-surface tanks discussed in the example above. These can also be used below pervious pavements to capture a proportion of the stormwater with the number of tanks selected in relation to demand with both unharvested and overflow water disposed of by infiltration or at a controlled rate to a surface water body or a combined sewer.

The most common application of PPS has been parking areas, whether they are extensive parking lots or individual house driveways and courtyards. The efficacy of pervious pavements in intercepting and immobilising or degrading traffic-derived pollutants has been the subject of a large body of research in many countries including England [14, 15], Scotland [16], Spain[17], Australia [18, 19] and the USA [20].

One of the most extensive experimental studies of PPS-derived water quality, with respect to irrigation, from pervious parking surfaces was instigated by Nnadi [21]. This was an indoor simulation experiment on a relatively large scale. The experimental outline was to apply oil drops (simulating oil dripping onto pavements from parked vehicles) and street dust to pervious pavement system models built into 1200 mm × 1000 mm × 600 mm heavy-duty polypropylene boxes (used for heavy industrial storage). Models were created with a selection of wearing courses including several cross-sections of new porous asphalt, a 10-year aged asphalt from a quarry parking surface, porous poured concrete and concrete paving blocks with infiltration channels. Thirteen millimetre simulated rainfall events were applied to the models at 15 mm/hour (52 min rain events) twice monthly. These were applied a day before and a day after the monthly application of used motor oil and dust. The effluents were captured and analysed for total suspended solids (TSS), metals and total petroleum hydrocarbons (TPH). The results were considered in relation to water quality as applied to irrigation and other forms of re-use as detailed in Nnadi et al. [2]. The results indicated that the porous asphalt and porous concrete are as good as the block paved permeable pavements in the treatment of stormwater pollution with all surfaces delivering water quality that is suitable for irrigation and clearly demonstrated the capability of the porous asphalt and concrete pavement system to trap hydrocarbons within the system.

While the use of pervious pavements for control of day-to-day automotive-derived pollutants is well established, it must be remembered that the traditional PPS has shown a poor pollution retention performance when challenged by major hydrocarbon releases [22]. Modification of the permeable pavement to incorporate a gravity separator just below the laying course has been reported for both aggregate-based [15] and plastic box-based pavements [23] showing a capability of retaining sufficient oil should every car in a parking lot simultaneously lose the content of their sump.

The first pot trials to investigate the acceptability of water quality from pervious pavements for irrigation originated at Coventry University as a PhD programme by one of the authors [21]. The source of water was a series of laboratory models. The models used were based on the modified designs by Newman [24] which incorporates a gravity separation system to ensure that free oil can never escape from the system and were artificially fertilized by adding Osmocote® slow release fertilizer and brushing it into the infiltration channels. This was to enhance the biodegradation rate of the hydrocarbons. The pot trial investigation showed that where the effluent water could be used for irrigation the excess nutrients provided distinct advantages to the growing plants [25]. In all cases, the plants irrigated with effluent grew much better than those irrigated with tap water. In effect, the system was being used for fertigation.

There has also been more recent work using a live car park, this case based on a highly modified design called a macro-pervious pavement system (MPPS) (**Figure 2**). In 2011, an investigation was initiated [26] on an MPPS installed as a prison car park in Scotland.

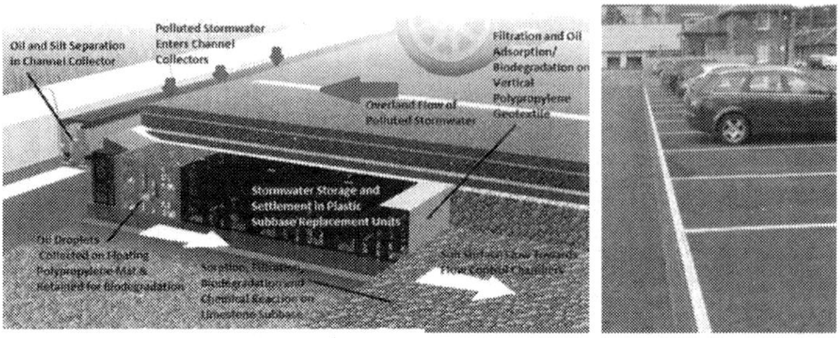

Figure 2. The macro-pervious pavement system as installed at Perth Prison, Scotland. The line of 'Channel Drain' is actually a series of oil and silt separating infiltration points, which direct stormwater into the sub surface storage and attenuation layer.

The parking lot consists of two major sub-catchments of around 1350 m^2 and a minor one of 300 m^2 (which was not included in the irrigation water study). The majority of the surface of the parking lot consists of traditional asphalt with stormwater entering the subbase/storage and attenuation layer of crushed limestone through miniaturized, linear, gravity separator units (which look like normal channel drains from the surface) through a chamber containing a floating mat of oil-sorbing textile. It then flows into the subbase, which drains towards separate flow control chambers (with orifice plate flow control) in each of the sub-catchments. These provided convenient sampling points and in-line storage chambers which would be available for landscape irrigation. Samples were analysed for a wide range of determinants.

For the heavy metals, all results were below the most stringent of irrigation water limits. For suspended solids, all measurements were below 20 mg/l, a concentration below which blockage of drip irrigation systems is not a problem. For TPH, the irrigation limits that were adopted are dependent on the solubility of the various carbon chain fractions (except for C_9–C_{14} where a limit of 1.8 mg/l applies). Since no measurements of TPH ever exceeded 1.8 mg/l and in the absence of any observed free product (indicating that solubility was never exceeded for any fraction), it can be concluded that hydrocarbons should have no detrimental effect if this water was harvested and used for irrigation.

BTEX concentrations were all below the 1 μg/l limits of detection for these compounds. The irrigation water limits for benzene and substituted benzenes proposed by the New Zealand Government [27] range from 800 μg/l for benzene to as high as 39 mg/l for toluene. MTBE was also invariably below its detection limit (10 μg/l). Irrigation water standards for MTBE could be found in the literature but since the State of Florida mandates a limit of 50 μg/l in drinking water, this value could be seen as a very conservative irrigation water limit. Boron was always below the 40 μg/l limit of detection and thus well below the 500 μg/l soil pore water limit for the most sensitive plants [28]. Examining the major cation concentrations, it was shown that calcium, magnesium and potassium were low compared to any irrigation water limits and most authors would expect that at least magnesium and potassium would need to be supplemented for optimum plant growth. For most of the year, neither the sodium (**Figure 6**) or sodium absorption ratio (SAR) values were a concern at this site, but in the winter months, following the application of de-icing salt (which will vary from year to year) the values of both these parameters show the water to be unsuitable for irrigation with sodium values in January up to 1200 mg/l. By March, the sodium concentrations were below 100 mg/l but above the 50 mg/l irrigation water limit until May. It was concluded that if water is to be stored for irrigation purposes it would be necessary to divert the meltwater (and the rainwater falling onto a salt contaminated pavement) away from the storage tanks during the winter months after filling them as much as possible during October and November and then topping them up in April and May before the need for irrigation is established in the summer. If the worst of the salt contaminated water can be diverted, the direct effect on plants will be minimized. Another issue from this particular site was the pH that was higher than optimal, but this was due to the limestone used in the subbase and would have been much lower if a granite subbase had been chosen. Adjustment of pH with addition of calcium sulphate was proposed as a potential solution, but species selection for high pH tolerance would be an alternative.

Pot trial experiments were carried out using both ryegrass and tomato plants (using the experimental protocol previously proposed by Nnadi [21] irrigated with effluent collected from this pavement in September with initial results being reported to the SuDSnet conference in 2015 [29] and continued experiments (and tissue analysis) have shown that at this site, the water is perfectly suitable for irrigation, if harvested before the salt application. However, the data represents only relatively short-term use of the water. Hence, further work in this area is required and this provides a good research opportunity for anyone interested in this area of study.

One problem with block paving surfaced PPS is that they can become a habitat for weeds and chemical methods of weed control still dominate as the most preferred in the United Kingdom [30] and most European cities [31, 32]. Although several herbicides are in use, glyphosate-containing herbicides (GCH) are by far the most widely used herbicide for weed control on hard surfaces [32–35]. Recent work on the impact of GCH on pollution attenuation and biodegradation in a PPS indicated that GCH is not retained in the PPS structure and is subject to rapid wash through in response to water movement thus increasing the potential risk of reaching receiving environments [36]. Hence, not only does one need to consider the herbicide itself, but the breakdown of other pollutants such as hydrocarbons within the PPS can be potentially affected [35] with potential impact on the quality of irrigation water. In a recent study to determine the suitability of stormwater harvested from PPS for reuse purposes in conditions where GCH was applied as part of PPS maintenance procedure, Mbanaso et al. [37] observed that effluent from the test models including those dosed with high GCH concentration of 7200 mg/l do not pose infiltration or salinity problems when used for irrigation. They, however, indicated that high dosage of the herbicide could lead to an elevated electrical conductivity of the recycled water. Hence, if PPS-derived water is used for irrigation; chemical methods of weed control should be avoided until the dynamics of breakdown of the particular herbicides are better understood. Further work needs to be done on establishing the lifetime of such compounds in both the PPS and the stored irrigation water.

3.1. Green roofs

As well as the need to control rapid water runoff in urban areas, there are several other problems to which SuDS can make a contribution. The urban heat island effect [38, 39] and reduction in green space with associated loss of both amenity and biodiversity [40] are important issues. Much of this is being exacerbated by climate change. A contribution to the mitigation of such problems is to apply green SuDS techniques and foremost amongst these is the attempt to utilise the roof spaces of buildings to create green roofs. Green roofs can be either extensive, usually planted with a mat of relatively drought-resistant species such as sedum or intensive [41], or being more 'garden like' where the aim goes beyond the immediate environmental benefits with aesthetic and social aims being contributed to by a more varied and attractive planting scheme with the potential to contribute more to biodiversity.

The potential of green roofs to contribute to the mitigation of rainwater runoff is well established, see Refs. [42–45]. What is less often stressed, however, are the problems in certain climates associated with the fact that after a heavy rainstorm, the water in the saturated substrate can take a considerable time to drain or evaporate and a subsequent storm will be offered significantly reduced attenuation.

The capability of green roofs to contribute to the cooling of buildings and combating the heat island effect has been widely reported [45–50]. Green roofs can also play a role in improving biodiversity within urban areas [40, 51] and, the aesthetic value of a green roof can often be a dominant factor in its adoption [52] even though aesthetic planting schemes can come into conflict with biodiversity aims [53].

Particularly where the green roof is intended to be accessible and provide, in part, the function, for example, of an urban park, it is important that the provision of water, to maintain adequate growing conditions (for a wide range of plants), during dry periods is recognised. This helps to enhance the amenity value. One of the factors is the maintenance of adequate soil moisture content. The limited load-bearing capacity (LBC) of the roofs of (existing) buildings often dictates the amount of water that can actually be safely retained on the roof itself. Designing traditional intensive green roofs to satisfy the load-bearing capacity of a roof thus involves finding a balance between storing as much rainwater as possible, maintaining conditions for plant growth and respecting the LBC of the construction [54, 55]. Adding more substrate typically achieves a water-stored-to-weight ratio (WSWR) of just 0.2 l of water per added kg of soil for loamy sand, or 0.4 l/kg for a typical, specialised, extensive green roof substrate [56].

An alternative to storing the water solely within the substrate is to drain the water into a cistern to allow the water to be used to irrigate the roof during dry periods and, if correctly sized, can offer temporary storage during storms occurring when the substrate is close to saturation. A green roof with a cistern for reuse was the subject of a modelling study by Hardin et al. [57]. Their system included a green roof with its drainage system connected to a cistern which in turn supplied irrigation water to the roof via a pump. A supplemental water source is also connected to the cistern to provide water should there not be sufficient water to perform the irrigation event. It was proposed that the irrigation should be managed via a controller, similar to what is widely used for home lawn irrigation, which only irrigates on the prescribed times unless sufficient rain has fallen within 24 h of the intended irrigation event.

Building rainwater cisterns inside or adjacent to the building and using pumps and irrigation systems is thus an option which will provide both runoff attenuation and on-site rainwater reuse. However, the running costs, capital costs and building space/land take combined with the need for maintenance and the propensity for active systems to break down are factors which would militate against their choice. There is also a need to consider the trade-off between energy used to pump the water from ground or basement level and the loss of space if the cistern is maintained in one of the higher floors of the building.

An alternative to using a separate cistern is to maintain 'ponding elements' under the substrate. Green roofs.com [58] presents a useful summary of some of the systems available, some of which are dependent on a slightly sloping roof, one which incorporates active pumped elements and a system that stores up to 40 mm of water in a plastic drainage layer equipped with an overflow device (although little extra capacity is available for temporary stormwater storage during rain events). However, without a separate cistern, the volume available for storage will be relatively small unless special steps are taken. One such approach has been developed in the Netherlands [56]. This modified green roof stores the water directly under the substrate within modified load bearing plastic void formers originally developed for pervious pavement applications. The water is not required to be pumped up to the substrate, this being achieved by capillary actions through 'capillary cones' inserted into the load-bearing vertical struts [59].

An excellent example of an application of this type of system is adjacent to the elevated railway station at Orlyplein in Amsterdam [56], shown in **Figure 3**. This is a remarkable installation which has transform the former rooftop bus station (vacant for some time after it being moved to ground level) into a popular public park area with a resulting increase in economic development into the form of many additional rooms constructed in adjacent hotels.

Figure 3. The roof park built on the former Orlyplein Bus Station *deck*, before, during and after construction.

The previously lonely and threatening exit from the rail station has been transformed into a place that is popular and well-populated whenever the station is open.

The system of capillary irrigation outlined above has also found application in ground level applications. A very recent example of this is the 'Green Stream' in Zuidas an Amsterdam city district in the Netherlands. Zuidas is a very densely build urban area and therefore prone to urban flooding during intense rain events. With the city district being under re-development, aiming to become the major business hub in the Amsterdam Metropolitan region, the city is developing innovative multifunctional designs to improve urban quality, reduce the urban heat-island effect by improved evaporative cooling by plants and increase water retention (and reuse) to prevent urban flooding. The 'Green Stream' is a project in which rainwater from rooftops and adjacent sidewalks is collected in a 2 m wide planting strip that runs along the houses and sidewalk (**Figure 4**). The strip is deeper than its surroundings, making it a natural water collection point in the street design. To get from the sidewalk to the houses, bridges are used to cross the 'green stream'. This planting strip is designed to be dry throughout the year, but is allowed to flood during rain events. Planting species are selected to be able to withstand occasional flooding. Innovative in the design is the 150 mm high water attenuation system 40 cm below the planting, which is fed with water from the roofs. A continuous chain of plastic void forming units, as used in the Orlyplein roof park, is placed in a waterproof liner to create a subsoil water-tank and features the same capillary irrigation system, capable of returning water to the soil when plants are using water without the use of pumps (and thus energy). The improved water availability for plants maintains their evapotranspiration rates at close to the potential evapotranspiration generated by the local weather, improving their urban cooling capacity. Surplus of water can drain freely to ground water level alongside the water drainage and capillary irrigation system. To prevent the Green Stream from overflowing onto the sidewalk, extra emergency overflows are created at the maximum fill level, connected to the conventional sewer.

Figure 4. Left to right: Schematic of the green stream system, under construction and on the day of completion.

Another application of this technology has been on sports surfaces such as football pitches both on rooftops and at ground level. The system provides water to the growing grass turf, while the void space can be made sufficiently deep to satisfy the most stringent stormwater attenuation requirements for new stadium construction. Even if supplemental water has to be

http://dx.doi.org/10.5772/67041

used to maintain the playing surface in very hot dry countries, the application of water from below by capillary action is more efficient than spray irrigation from above. Currently, trials are underway in preparation for the soccer world cup in Qatar.

3.2. Irrigation of urban trees

Novel below ground water storage options provide a significant opportunity for the reliable delivery of acceptable water to trees in the urban environment. In a non-urban environment and under natural establishment conditions, a mature tree have grown partly due to a favourable water regime that has provided sufficient water reliably over the life of the tree. Historically, urban trees have faced significant challenges that are not faced by non-urban trees which include:

- Establishment in unfavourable environments
- A deficiency in total water volume provision
- Runoff of water from non-permeable surfaces away from tree roots, even where local rainfall totals are adequate
- Low retention rates of applied water, where water evaporates or percolates away from tree roots before uptake
- Insufficient mature and productive soil/substrate for nutrient regeneration and incomplete establishment of beneficial microbial processes (e.g. limited growth of mycorrhizae)
- Tree removal or inappropriate management resulting in damage to trees if land use changes regard existing trees as an obstacle

Thus, water supply considerations provide some of the major challenges faced by urban trees and depending on the tree species, a mature tree with a 500 mm trunk diameter could require 860l of water to be provided for survival and normal function during drought conditions. It is important to acknowledge the local hydrological benefits of urban trees in times of intense and or prolonged rainfall where the drainage system may be at capacity. The positive impacts of trees include rainfall detention, retention and uptake, which may remove significant volumes of water from runoff totals, or delay local peak discharge in comparison with a location without trees [60].

In theory, it would be acceptable to deliberately divert stormwater from a new development to supplement existing urban trees to meet the total water need. This could include disconnection of downpipes into soil and conveyance to areas where trees are located. However, this retrofit option could result in standing water in the case of intense events if infiltration and percolation are not sufficient and a bypassing of the tree if the water is not retained by the soil. Hence, solutions are required that provide the required volume of water, but allow this to be retained close to the roots and provide useful water. In practice, this is necessarily targeted at new developments and is usually provided by tree pits, where a selected soil medium is placed into a structural chamber. Landscaping solutions can incorporate this type of structure into bioretention schemes as shown by the image below of a parking area for a shopping Centre drained by bioretention cells, which often successfully incorporate trees (**Figure 5**).

Figure 5. A bioretention cell with trees and shrubs in Raleigh, North Carolina, USA.

The medium can be an engineered soil or a replacement medium such as absorbent foam-filled structurally resistant void forming boxes. An ultra-absorbent foam that is available commercially as a component in a tree box system was tested independently in a laboratory for its impact on hydraulic conditions and was shown to be a promising solution for runoff management and water absorption [61]. This material also has the potential for water transfer to woody plants.

The use of engineered materials, including allochthonous soils, in providing water to trees has not been proven to fully replicate the functioning of the biological component of mature natural soils, nor have an equivalent of the nutrient recycling and regeneration of soil. However, it has been established that drainage systems that include trees and soil can improve water quality in sustainable drainage applications, including the absorption and retention of heavy metals, nitrogen and phosphorus [62]. The addition of urban pollutants in the form of sediment moved by stormwater is shown in **Figure 6**. In such circumstances, however, the blinding of the surface can become an issue and a means of pre-filtration or sedimentation to take

Figure 6. A tree pit receiving urban sediment from an adjacent parking area, Greensborough, North Carolina, USA.

http://dx.doi.org/10.5772/67041

out the worst of the sediment may be advisable and if not available close control of maintenance may be required.

The understanding of the requirements for successful irrigation of urban trees has improved in recent years, particularly as the recognition of the full range of benefits provided by urban trees has become apparent and the use of trees in the urban landscape has been more readily encouraged. As shown in the plates in this section, care must be taken to prevent waterlogging of supporting soils, partly due to the confined nature of tree pits and also the provision of water in excess from large impermeable areas, which may exclude the air that is found in natural soils and may, if serious enough, lead to anoxic conditions and the production of greenhouse gases. Where possible, water should be directed to a sub-root zone position without passing through the root ball itself. Consideration of aeration and an under drain in design should prevent this, alongside a consideration of local rainfall totals and the nearby landscape conditions.

4. Conclusion

SuDS elements, by their nature, store water but it is not always simple to either make a SuDS system provide a suitable supply of irrigation water or to make a water harvesting system contribute to stormwater source control. Both quality issues and quantity issues are important and in areas where road salt is used ground level collection needs to be carefully managed to prevent stored water becoming heavily contaminated with salt that could make the water unsuitable for irrigation and the collection mechanism needs to provide attenuation of automobile-derived pollutants sufficient to allow plants to grow without inhibition. Where the prime aim is to harvest a good quality of water such as on a roof, whether this is used directly to irrigate plants on the roof or is collected for offline irrigation sizing of tanks needs to allow enough excess storage to deal with storm events and the tanks need to be managed to ensure such volume is routinely available.

Author details

Andrew B. Shuttleworth[1], Ernest O. Nnadi[2*], Fredrick U. Mbanaso[2], Stephen J. Coupe[3], Joris G.W.F. Voeten[4] and Alan P. Newman[2]

*Address all correspondence to: dr.nnadi@gmail.com

1 SEL Environmental Ltd, Blackburn, Lancashire, United Kingdom

2 School of Energy, Construction and Environment, Coventry University, Coventry, United Kingdom

3 Centre for Agroecology, Water and Resilience, Coventry University, Coventry, United Kingdom

4 Permavoid Ltd, Amsterdam, Netherlands

References

[1] FAO, (2008), Wastewater Quality Guidelines for Agricultural Use. Food and Agricultural Organisation Corporate Document Repository: Natural Resources Management and Environmental Department [online]. http://www.fao.org/docrep/T0551E/t0551e04.htm <25 September 2016>

[2] Nnadi, E.O., Coupe, S.J., Newman, A.P. and Mbanaso, F.U., (2015) 'Stormwater harvesting for irrigation purposes: an investigation of chemical quality of water recycled in pervious pavement system' Journal Environmental Management, 147, 246–256.

[3] Meera V. and Mansoor Ahammed M. (2006) Water quality of rooftop rainwater harvesting systems: a review, Journal of Water Supply: Research & Technology, 55 (4) 257–268.

[4] Jones, M.P. and Hunt, W. F. (2008) Rainwater Harvesting (2008) 'Urban Waterways: Guidance for Homeowners', North Carolina Cooperative Extension Service AGW-588-11. https://content.ces.ncsu.edu/rainwater-harvesting-guidance-for-homeowners

[5] DeBusk, K.M. and Hunt, W.F. (2014) 'Rainwater Harvesting: a Comprehensive Review of Literature', Report No. 425 NC Water Resources Research Institute. http://repository.lib.ncsu.edu/dr/bitstream/1840.4/8170/3/NC-WRRI-425.pdf

[6] DeBusk, K.M, Hunt, W.F, Osmond, D.L and Cope, G.W. (n.d.) 'Water Quality of Rooftop Runoff: Implications for Residential Water Harvesting Systems', North Carolina Cooperative Extension, https://content.ces.ncsu.edu/water-quality-of-rooftop-runoff

[7] Booth, D. B. and Leavitt, J. (1999), 'Field evaluation of permeable pavement systems for improved stormwater management', Journal of the American Planning Association, 65(3), 314–325.

[8] Gomez-Ullate, E., Bayon, J. R., Coupe, S. and Castro-Fresno, D. (2010). 'Performance of pervious pavement parking bays storing rainwater in the north of Spain', Water Science and Technology. 62(3), 615–621.

[9] Klein, R. D. (1979), 'Urbanization and stream quality impairment' Water Resources Bulletin, 15(4), 948–963.

[10] Rushton, B. T. (2001), 'Low-impact parking lot design reduces runoff and pollutant loads', Journal of Water Resources Planning and Management-ASCE, 127(3), 172–179.

[11] Sañudo-Fontaneda, L.A., Charlesworth, S., Castro-Fresno, D. andrés-Valeri, V.C.A. and Rodriguez-Hernandez J. (2014) 'Water quality and quantity assessment of pervious pavements performance in experimental car park areas', Water Science and Technology, 69(7), 1526–1533.

[12] Scholz, M. and Grabowiecki, P. (2007), 'Review of permeable pavement systems', Building and Environment, 42(11), 3830–3836.

http://dx.doi.org/10.5772/67041

[13] Pratt, C.J. (1996) 'Permeable Pavements for Stormwater Quality Enhancement' Proceeding Paper Urban Stormwater Quality Enhancement: Source Control, Retrofitting and Combined Sewer Technology American Society of Civil Engineers.

[14] Newman, A. P., Pratt, C. J., Coupe, S. J. and Cresswell, N. (2002), 'Oil bio-degradation in permeable pavements by microbial communities', Water Science and Technology, 45(7),51–56.

[15] Newman, A., Puehmeier, T., Shuttleworth, A. and Pratt, C.J. (2014) Performance of an enhanced pervious pavement system loaded with large volumes of hydrocarbons. Water Science and Technology, 70(5), 835–842.

[16] Macdonald, K. and Jefferies, C. (2001), 'Performance Comparison of Porous Paved and Traditional Car Parks' Proceedings of First National Conference on Sustainable Drainage, Coventry University, ISBN 1 903818 06 0, pp. 170–181.

[17] Bayón J.R., Castro D., Moreno-Ventas X., Coupe S.J. and Newman A.P., Pervious pavement research in Spain: Hydrocarbon degrading microorganisms, Proceedings of 10th International Conference on Urban Drainage, Copenhagen/Denmark, 21–26 August 2005. CD-ROM

[18] Jayasuriya, N., Kadurupokune, N., Othman, M. and Jesse, K. (2007) 'Contributing to the sustainable use of stormwater: the role of pervious pavements' Water Science & Technology, 56 (12), 69–75.

[19] Fletcher, T.D., Deletic, A., Mitchell, V.G. and Hatt, B.E (2008) 'Reuse of urban runoff in Australia: a review of recent advances and remaining challenges', Journal of Environmental Quality, 37(5), 116–127.

[20] Brattebo, B. O. and Booth, D. B. (2003), 'Long-term stormwater quantity and quality performance of permeable pavement systems', Water Research, 37(18), 4369–4379.

[21] Nnadi, E.O. (2009) 'An evaluation of modified pervious pavements for water harvesting for irrigation purposes' Unpublished PhD Thesis. Coventry: Coventry University.

[22] Newman, A. P., Puehmeier, T., Kwok, V., Lam, M., Coupe, S. J., Shuttleworth, A. and Pratt, C. J. (2004), Protecting groundwater with oil-retaining pervious pavements: Historical perspectives, limitations and recent developments, Quarterly Journal of Engineering Geology and Hydrogeology, 37(4), 283–291.

[23] Wilson, S., Newman, A.P., Puehmeier, T. and Shuttleworth, A. (2003). Performance of an oil interceptor incorporated into a pervious pavement, ICE Proc: Engineering Sustainability, 156(1), 51–58.

[24] Newman, A. P. (2003), Pollutant containment system, WO03040483 [patent]

[25] Nnadi, E.O., Newman, A.P. and Coupe, S.J., (2014), 'Geotextile incorporated permeable pavement system as potential source of irrigation water: effects of re-used water on the soil, plant growth and development', Clean-Soil Air Water, 42(2), 125–132.

[26] Newman, A.P., Aitken, D. and Antizar-Ladislao, B. (2013), Stormwater quality performance of a macro-pervious pavement car park installation equipped with channel drain based oil and silt retention devices, Water Research, 47(2), 7327–7336.

[27] New Zealand Govt. (2011), Guidelines for Assessing and Managing Petroleum Hydrocarbon Contaminated Sites [online]. http://mfe.govt.nz/publications/hazardous/oil-guide-jun99/module-5.pdf

[28] Bauder, T.A., Waskom, P.L., Sutherland, P.L. and Davis, J.G. (2014) 'Irrigation Water Quality Criteria' Factsheet 0.506 Colorado State University: http://www.ext.colostate.edu/pubs/crops/00506.html

[29] Newman A.P., Nnadi, E.O., Mbanaso, F.U., Shuttleworth, A., Aitken D., Antizar-Ladislao B. and Theophilus S.C. (2015) 'Potential of Pervious and MacroPervious Pavements as Harvesting Systems for Irrigation of Adjacent Lawns and Flower Beds' SUDSnet International Conference 2015 3rd & 4th September 2015 Techno-Centre, Coventry University, UK. http://sudsnet.abertay.ac.uk/documents/SUDSnet2015_Newmanetal_PotentialofPerviousandMacroPerviousPavementsasHarvestingSystemsfo_000.pdf

[30] Tuffnell, N. and Britt, C. (2008) Determining the Usage and Usage Patterns of Amenity Pesticides across the UK (PS2230) [online]. Available from http://www.pesticides.gov.uk/uploadedfiles/Web_Assets/PSD/RPA_Amenity_Report_2008.pdf [08/07/2016]

[31] Rask, A.M. (2012) Non-chemical weed control on hard surfaces: An investigation of long-term effects of thermal weed control methods. Forest & Landscape Research No. 52-2012. Forest & Landscape Denmark, Frederiksberg. 156.

[32] Kempenaar, C. (2010) Best practice for chemical weed control on hard surfaces. European Glyphosate Environmental Information Source, Wageningen: Wageningen University and Research Centre - Plant Research International.

[33] Kolpin, D.W., Thurman, E.M., Lee, E.A., Meyer, M.T., Furlong, E.T. and Glassmeyer, S.T. (2006), 'Urban contributions of glyphosate and its degradate AMPA to streams in the United States', Science of the Total Environment, 354, 191–197.

[34] Kristoffersen, P., Rask, A.M., Grundy, A.C., Franzen, I., Kempenaar, C., Raisio, J., Schroeder, H., Spijker, J., Verschwele, A. and Zarina, L. (2008), 'A review of pesticide policies and regulations for urban Amenity areas in seven European countries', Weed Research 48(3), 201–214.

[35] Woodburn, A.T., (2000), 'Glyphosate: production, pricing and use worldwide', Pest Management Science, 56, 309–312.

[36] Mbanaso, F.U., Coupe, S.J., Charlesworth, S.M. and Nnadi, E.O. (2012), Laboratory-based experiments to investigate the impact of glyphosate-containing herbicide on pollution attenuation and biodegradation in a model pervious paving system, Chemosphere, 90 (2013), 737–746.

[37] Mbanaso, F.U., Nnadi, E.O., Coupe, S.J. and Charlesworth, S.M. (2016), Stormwater harvesting from landscaped areas: effect of herbicide application on water quality and usage, Environmental Science and Pollution Research, 23, 15970.

[38] Baik, J., Kwak, K., Park, S. and Ryu, Y. (2012), 'Effects of building roof greening on air quality in street canyons', Atmospheric Environment, 61, 48–55.

[39] Bowler, D.E., Buyung-Ali, L., Knight, T.M. and Pullin, A.S. (2010). Urban greening to cool towns and cities: a systematic review of the empirical evidence. Landscape and Urban Planning, 97, 147–155.

[40] Brenneisen S., (2006), Space for urban wildlife: designing green roofs as habitats in Switzerland, Urban Habitats, 4(1), 27–36.

[41] Hunt, W.F. and Szpir, L.L. (n.d.,) Permeable Pavements, Green Roofs and Cisterns Stormwater Treatment Practices for Low-Impact Development, Pub. North Carolina Cooperative Extension Service https://www.bae.ncsu.edu/extension/ext-publications/water/protecting/ag-588-06-permeable-pavements-green-roofs-and-cisterns.pdf

[42] Stovin V. (2010), 'The potential of green roofs to manage urban stormwater', Water and Environment Journal, 24(3), 192–199.

[43] Carson, T.B., Marasco, D.E., Culligan, P.J. and McGillis, W.R.(2013), 'Hydrological performance of extensive green roofs in New York City: observations and multi-year modeling of three full-scale systems', Environmental Research Letters, 8, online at; http://iopscience.iop.org/article/10.1088/1748-9326/8/2/024036/pdf

[44] Fioretti, R., Palla, A., Lanza, L.G. and Principi, P. (2010) Green roof energy and water related performance in the Mediterranean climate, Building and Environment, 45, 1890–1904.

[45] Wong N.H. , Chen Y., Ong C.L. and Sia A. (2003), Investigation of thermal benefits of rooftop garden in the tropical environment, Building and Environment, 38(2), 261–270.

[46] Palomo Del Barrio, E.(1998) ,Analysis of the green roofs cooling potential in buildings, Energy and Buildings, 27(2), 179–193.

[47] Castleton, H.F., Stovin, V., Beck, S.B.M. and Davison, J.B. (2010), 'Green roofs: building energy savings and the potential for retrofit', Energy and Buildings, 42(10), 1582–1591.

[48] Alexandri, E. and Jones P.(2008), Temperature decreases in an urban canyon due to green walls and green roofs in diverse climates, Building and Environment, 43(4), 480–493.

[49] Takebayashi, H. and Moriyama, M.(2007), Surface heat budget on green roof and high reflection roof for mitigation of urban heat island, Building and Environment, 42(8), 2791–2979.

[50] Susca T., Gaffin S.R. and Dell'Osso G.R.(2011), Positive effects of vegetation: urban heat island and green roofs, Environmental Pollution, 159(8–9), 2119–2126.

[51] Kadas G. (2006), 'Rare invertebrates colonizing green roofs in London', Urban Habitats, 4, 66–86.

[52] Matsuoka, R. H. and Kaplan, R. (2008), "People needs in the urban landscape: analysis of landscape and urban planning contributions", Landscape Urban Planning, 84(1), 7–19.

[53] Dunnett, N. (2006) 'Green Roofs for Biodiversity: Reconciling Aesthetics with Ecology', Proc. Fourth Annual Greening Rooftops for Sustainable Communities Conference, Awards and Trade Show Proceedings May 10–12, 2006, Boston, MA.

[54] Molineux, C.J., Fentiman, C.H. and Gange, A.C.(2009), Characterising alternative recycled waste materials for use as green roof growing media in the UK, Ecological Engineering, 35, 1507–1513.

[55] Snodgrass, E. and McIntyre, L.(2010), The Green Roof Manual, Portland, OR: Timber Press.

[56] Voeten J.W.G.F., van de Werken L. and Newman A.P. (2016) Demonstrating the Use of Below-Substrate Water Storage as a Means of Maintaining Green Roofs - Performance Data and a Novel Approach to Achieving Public Understanding, Proc. ;World Environmental and Water Resources Congress 2016, Professional Development, Innovative Technology, International Perspectives and History and Heritage, Pathak CS and Reinhart D (Eds) American Society of Civil Engineers , Reston.

[57] Hardin, M., Wanielista, M. and Chopra, M. (2012) 'A mass balance model for designing green roof systems that incorporate a cistern for re-use', Water, 4, 914–931.

[58] Greenroofs.com (n.d.) Water Storage & Irrigation. http://www.greenroofs.com/Greenroofs101/water-storage.htm

[59] Van Raam C.H. Shuttleworth A.B. and Culleton P.D. (2014) Plant surface structure and modules and method for forming the same, Patent. WO2014006180 A1

[60] Stovin, V.R., Jorgensen, A. and Clayden, A. (2008) 'Street trees and stormwater management' Arboricultural Journal, 30(4), 297–310.

[61] Nnadi, E.O., Coupe, S.J., Sañudo-Fontaneda, L.A. and Rodriguez-Hernandez, J. (2014), 'An evaluation of enhanced geotextile layer in permeable pavement to improve stormwater infiltration and attenuation', International Journal of Pavement Engineering, 15(10), 925–932.

[62] Hunt, W.F., Smith, J.T., Jadlocki, S.J., Hathaway, J.M. and Eubanks, P. R. (2008) Pollutant removal and peak flow mitigation by a bioretention cell in urban Charlotte, N.C., Journal of Environmental Engineering, 134(5), 403–408.